Geographic Data Science with R

The burgeoning field of data science has provided a wealth of techniques for analysing large and complex geospatial datasets, including descriptive, explanatory, and predictive analytics. However, applying these methods is just one part of the overall process of geographic data science. Other critical steps include screening for suspect data values, handling missing data, harmonizing data from multiple sources, summarizing the data, and visualizing data and analysis results. Although there are many books available on statistical and machine learning methods, few encompass the broader topic of scientific workflows for geospatial data processing and analysis.

The purpose of *Geographic Data Science with R* is to fill this gap by providing a series of tutorials aimed at teaching good practices for using geospatial data to address problems in environmental geography. It is based on the R language and environment, which currently provides the best option for working with diverse spatial and non-spatial data in a single platform. Fundamental techniques for processing and visualizing tabular, vector, and raster data are introduced through a series of practical examples followed by case studies that combine multiple types of data to address more complex problems.

The book will have a broad audience. Both students and professionals can use it as a workbook to learn high-level techniques for geospatial data processing and analysis with R. It is also suitable as a textbook. Although not intended to provide a comprehensive introduction to R, it is designed to be accessible to readers who have at least some knowledge of coding but little to no experience with R.

Key Features:

- Focus on developing practical workflows for processing and integrating multiple sources of geospatial data in R
- Example-based approach that teaches R programming and data science concepts through real-world applications related to climate, land cover and land use, and natural hazards.
- Consistent use of tidyverse packages for tabular data manipulation and visualization.
- Strong focus on analysing continuous and categorical raster datasets using the new terra package
- Organized so that each chapter builds on the topics and techniques covered in the preceding chapters
- Can be used for self-study or as the textbook for a geospatial science course.

CHAPMAN & HALL/CRC DATA SCIENCE SERIES

Reflecting the interdisciplinary nature of the field, this book series brings together researchers, practitioners, and instructors from statistics, computer science, machine learning, and analytics. The series will publish cutting-edge research, industry applications, and textbooks in data science.

The inclusion of concrete examples, applications, and methods is highly encouraged. The scope of the series includes titles in the areas of machine learning, pattern recognition, predictive analytics, business analytics, Big Data, visualization, programming, software, learning analytics, data wrangling, interactive graphics, and reproducible research.

Published Titles

Statistical Foundations of Data Science
Jianqing Fan, Runze Li, Cun-Hui Zhang, and Hui Zou

A Tour of Data Science: Learn R and Python in Parallel
Nailong Zhang

Explanatory Model Analysis
Explore, Explain, and, Examine Predictive Models
Przemyslaw Biecek, Tomasz Burzykowski

An Introduction to IoT Analytics
Harry G. Perros

Data Analytics
A Small Data Approach
Shuai Huang and Houtao Deng

Public Policy Analytics
Code and Context for Data Science in Government
Ken Steif

Supervised Machine Learning for Text Analysis in R
Emil Hvitfeldt and Julia Silge

Massive Graph Analytics
Edited by David Bader

Data Science
An Introduction
Tiffany-Anne Timbers, Trevor Campbell and Melissa Lee

Tree-Based Methods
A Practical Introduction with Applications in R
Brandon M. Greenwell

Urban Informatics
Using Big Data to Understand and Serve Communities
Daniel T. O'Brien

Introduction to Environmental Data Science
Jerry Douglas Davis

Hands-On Data Science for Librarians
Sarah Lin and Dorris Scott

For more information about this series, please visit: https://www.routledge.com/Chapman--Hall-CRC-Data-Science-Series/book-series/CHDSS

Geographic Data Science with R

Visualizing and Analyzing Environmental Change

Michael C. Wimberly

CRC Press
Taylor & Francis Group
Boca Raton London New York

CRC Press is an imprint of the
Taylor & Francis Group, an **informa** business

A CHAPMAN & HALL BOOK

Designed cover image: © Michael C. Wimberly

First edition published 2023
by CRC Press
6000 Broken Sound Parkway NW, Suite 300, Boca Raton, FL 33487-2742

and by CRC Press
4 Park Square, Milton Park, Abingdon, Oxon, OX14 4RN

CRC Press is an imprint of Taylor & Francis Group, LLC

ISBN: 978-1-032-34771-4 (hbk)
ISBN: 978-1-032-35288-6 (pbk)
ISBN: 978-1-003-32619-9 (ebk)

DOI: 10.1201/9781003326199

Typeset in Latin modern
by KnowledgeWorks Global Ltd.

Publisher's note: This book has been prepared from camera-ready copy provided by the authors.

Access the companion website: insert CW URL

To my parents,

who taught me to value learning and supported my academic endeavors

Contents

List of Figures

List of Tables

Preface

We live in a time of unprecedented environmental change, driven by the effects of fossil fuels on the Earth's climate and the expanding footprint of human land use. To mitigate and adapt to these changes, there is a need to understand their myriad impacts on human and natural systems. Achieving this goal requires geospatial data on a variety of environmental factors, including climate, vegetation, biodiversity, soils, terrain, water, and human populations. Consistent monitoring is also necessary to identify where changes are occuring and determine their rates. Large volumes of relevant data are collected by Earth-observing satellites and ground-based sensors. However, the data alone are not enough. Using them effectively requires tools for appropriate manipulation and analysis.

The burgeoning field of data science has provided a wealth of techniques for analyzing large and complex datasets, including methods for descriptive, explanatory, and predictive analytics. However, actually applying these methods is typically a small part of the overall data science workflow. Other critical tasks include screening for suspect data, handling missing values, harmonizing data from multiple sources, summarizing variables for analysis, and visualizing data and analysis results. Although there are now many books available on statistical and machine learning methods, there are fewer that address the broader topic of scientific workflows for geospatial data processing and analysis.

The purpose of *Geographic Data Science with R* (GDSWR) is to fill this gap. GDSWR provides a series of tutorials aimed at teaching good practices for using time series and geospatial data to address topics related to environmental change. It is based on the R language and environment, which currently provides the best option for working with diverse sources of spatial and non-spatial data using a single platform. The book is not intended to provide a comprehensive overview of R. Instead, it uses an example-based approach to present practical approaches for working on diverse problems using a variety of datasets.

The material in GDSWR was originally developed for upper-level undergraduate and graduate courses in geospatial data science. It is also suitable for individual study by students or professionals who want to expand their capabilities for working with geospatial data in R. Although the book is not intended to be a comprehensive reference manual, it can also be useful for readers who are looking for examples of particular methods that can be modified for new applications. The tutorials focus on physical geography and draw upon a variety of data sources, including weather station data, gridded climate data, classified land cover data, and digital elevation models. It is my sincere hope that GDSWR will help readers increase their proficiency with R so that they can implement more sophisticated data science workflows that make effective use of diverse geographic data sources. These skills will allow them to address pressing scientific questions and develop new geospatial applications that can enhance our understanding of the changing world we inhabit.

Rationale and Philosophy

GSDWR is aimed at readers who have already taken one or more introductory courses in Geographic Information Sciences (GIS) or who have experience working with GIS software and geospatial data. It assumes that readers are familiar with basic geospatial data structures, such as vector and raster data, along with basic cartographic concepts such as projections and coordinate systems. Readers who have a limited background in GIS and other geospatial technologies should consider reviewing an introductory text such as Paul Bolstad's *GIS Fundamentals* (Bolstad, 2019).

The R language and environment for statistical computing is one of the most widely used platforms for data analysis. Two of the major advantages of R are that it is available to users at no cost and that it can be used with a multitude of add-on packages that provide customized methods for working with particular types of data or applying specialized analytical tools. However, this breadth and complexity also present a significant challenge for users who want to learn R and use it to analyze their own data. It is frequently remarked that anything that can be done in R can be accomplished in multiple ways. This flexibility can be a strength in the hands of an experienced user, but to learners, it is often overwhelming and baffling. These observations have strongly influenced the design of my university classes and workshops that incorporate R, and they inform the structure and content of *Geographic Data Science with R* (GSDWR).

The book is not intended to be a comprehensive introduction to R, but it provides a path for new users to quickly master basic skills and move on to more sophisticated data analysis. To this end, GDSWR begins with an introductory chapter that highlights key aspects of the base R language that provide a foundation for working with more specialized packages in the later chapters. Readers who want more extensive knowledge of R should consult one of the many introductory texts that are available, including Tilman Davies' *The Book of R* (Davies, 2016), and Rob Kabacoff's *R in Action* (Kabacoff, 2015).

GDSWR introduces several major R packages that facilitate the processing and analysis of tabular data along with vector and raster forms of geospatial data. In particular, the book makes extensive use of the **tidyverse** collection of R packages. More background on the general concepts of "tidy" data organization and the specific tidyverse packages and functions can be found in Wickham and Grolemund's *R For Data Science* (Wickham and Grolemund, 2016). The packages used in this book have been selected because they provide powerful and generalizable frameworks for data processing, visualization, and analysis. They are well maintained, have a broad user base, and are useful for a variety of data science applications. In many cases, additional functions from other packages could be incorporated to make particular tasks more convenient or efficient. However, the overall approach of GDSWR is to focus on a more limited set of powerful tools for students to master and eventually apply to a much broader set of problems.

To the extent possible, GSDWR has been organized so that the techniques learned in each chapter continue to be applied and expanded upon in the subsequent chapters. Chapter 1 provides a brief overview of important concepts in R that can serve as an introduction for readers who have not worked with R before or as a refresher for more experienced users. Chapter 2 then introduces the **ggplot2** package (Wickham et al., 2022a) for scientific graphs. This package is used throughout the book to generate charts and maps. Chapter 3 introduces the **dplyr** and **tidyr** packages (Wickham et al., 2022b; Wickham and Girlich, 2022) for the manipulation of non-spatial data frames. Chapter 4 provides an overview on how to work with dates using the **lubridate** package (Spinu et al., 2021). These techniques are then applied to geospatial data in Chapter 5, which introduces the **sf** package (Pebesma, 2022) for storing and processing spatial features and attributes. Chapters 6 and 7 introduce the *terra* package (Hijmans, 2022) for manipulating and analyzing raster data, and Chapter 8 provides an overview of tools and approaches for integrating geographic datasets with different coordinate reference systems. Chapters 9 and 10 present examples of more complex analyses that combine multiple vector and raster datasets. Finally, Chapters 11 and 12 present examples of specific applications of the techniques covered in the book for analyzing patterns of wildfire severity and modeling species ranges with ecological niche models and climate data.

How to Use this Book

If you are new to R, it is best to start from the beginning and progress through the chapters sequentially, as each one builds on topics covered previously. More experienced R users, particularly those who have already worked with the packages in the **tidyverse** collection (Wickham, 2021), should be able to skip to specific chapters of interest. Each chapter provides narrative text along with blocks of R code and the outputs from running the code. In the text, package names are in bold text (e.g., **ggplot2**), and inline code, function arguments, object classes, object names, and filenames are formatted in a typewriter font (e.g., `myobject`). Function names are followed by parentheses (e.g., `ggplot()`). There are many excellent learning resources and reference guides that expand on the topics covered in GDSWR. Recommendations for further study are provided at appropriate locations throughout the book.

The data files used in each chapter can be downloaded from `https://doi.org/10.6084/m9.figshare.21301212` and used to run the code on your own system. One of the best ways to learn programming is to experiment by modifying existing code to see how the outputs change as a result. At the end of each chapter, there are several suggestions for how to practice by modifying the example code. When running the code, we recommend that readers use the RStudio graphical user interface (GUI) software and create a separate project for each chapter. The code for file input and output assumes that data will be read from and written to the working directory of the RStudio project. The appendix includes instructions on how to set up RStudio projects and associate them with folders in your computer's file system.

Finally, readers should not hesitate to use the techniques herein for their own projects with new datasets. For example, simple data manipulation and graphing tasks that are done with a spreadsheet can be automated with R scripts. Spatial analyses that are done interactively with dedicated GIS software can similarly be translated into R code. Once various analysis steps are implemented in R, it is much easier to combine them into well-documented reproducible workflows. Making the transition from tutorials to designing analyses and writing your own code will require considerable trial and error. As with any endeavor, practice and perseverance are ultimately the keys to learning to use R effectively for geographic data science.

Acknowledgments

My intellectual growth as a scientist and my capabilities as a data analyst have benefited from the many fabulous colleagues with whom I have had the privilege to collaborate. In particular, I owe a debt of gratitude to all the students, postdocs, and research staff who have worked with me over the years. Furthermore, my development as a teacher has been strongly influenced by the many students who have taken my university courses and professional workshops. The approach to geospatial data science presented in this book represents many years of experimentation and refinement in response to their suggestions and feedback. Finally, this book would not have been possible without the unwavering support of my wonderful family. Thank you Anne, Alice, Zach, Slushie, Ivo, Sadie, and Pepper.

Mike Wimberly,
Norman, OK

About the Author

Dr. Michael Wimberly is a Professor in the department of geography and environmental sustainability at the University of Oklahoma. He previously served on the faculties of the University of Georgia and South Dakota State University. He holds a PhD from Oregon State University, an MS from the University of Washington, and a BA from the University of Virginia. Dr. Wimberly has taught graduate and undergraduate courses in Geographic Information Science (GIS), Spatial Statistics, and Landscape Ecology. He has worked with the R language and software environment throughout his academic career after being introduced to its predecessors, S and S+, in graduate school. His research combines ecological models with Earth observation data to address scientific questions and create practical applications in the fields of public health and natural resource management. Areas of study include the effects of land use and climate on vector-borne disease transmission, wildfire and vegetation dynamics in temperate and tropical forest ecosystems, and the causes and consequences of agricultural expansion into native forests and grasslands.

1

Introduction to R

This chapter provides a brief introduction to the R programming language. The goal is to give just enough background to allow readers to move on quickly to the subsequent chapters, which focus on more advanced applications of R for geospatial data processing, analysis, and visualization. To understand and apply these techniques, it is essential to know the basic R *objects* that store various types of data. It is also necessary to understand how *functions* are used to manipulate these objects. Functions typically take one or more objects as inputs and modify them to produce a new object as the output. These concepts will be demonstrated by using R to perform a series of simple data processing tasks such as making basic calculations, applying these calculations over larger datasets, querying subsets of data that meet one or more conditions, creating graphics, and carrying out basic statistical tests.

1.1 Basic Calculations

The simplest way to use R is to type in a mathematical expression at the command prompt in the console and hit the Enter key. R will calculate the answer and print it on the screen. Alternatively, you can type the mathematical expressions as lines in an R script file. The code can then be run one line at a time using Ctrl+Enter on the keyboard or the Run button in the RStudio interface. Multiple lines can be run by highlighting them in the R script file and then using Ctrl+Enter or the Run button.

```
23 + 46
## [1] 69
27 * 0.003
## [1] 0.081
160 + 13 * 2.1
## [1] 187.3
(160 + 13) * 2.1
## [1] 363.3
```

It is strongly recommended that you work with R scripts from the outset rather than entering code directly into the console. Writing your code in a script file makes it easy to see the lines that have already been run, detect and correct any errors, and save the code so that you can share it or continue to work on it later. Eventually, you will develop more complex scripts with many lines of code to implement complex workflows. In RStudio, you can create a new script file by selecting File > New File > R Script from the RStudio menu or using the Ctrl+Shift+N key combination. Script files can be saved and opened using the File menu in RStudio or with keyboard shortcuts such as Ctrl+O to open and Ctrl+S to save.

Instead of just printing the results on the screen, the outputs of mathematical expressions can be saved as variables using the leftward assignment operator (<-). Creating variables allows information to be stored and reused at a later time. R stores each variable as an *object*.

```
x <- 15
x
## [1] 15
y <- 23 + 15 / 2
y
## [1] 30.5
z <- x + y
z
## [1] 45.5
```

A single equal sign, =, can also be used as an assignment operator in R like in many other programming languages. This book exclusively uses the <- operator for clarity. The single equal sign is also used in R to associate function arguments with values, which can lead to ambiguity and confusion. It is also easy to confuse the = assignment operator with the == logical operator.

When an object name is entered at the command line or through a script, R will invoke the print() function and output information about the object to the console. The print() function is an example of a generic function that can take a variety of object classes as input. Each class has an associated *method* that determines how it will be printed. Thus, a generic function like print() will produce different results depending on the characteristics of the input.

The R console is used to get "quick looks" at the data and analysis results. However, the text format of the console greatly limits what can be displayed. In addition to printing results to the screen, they can also be exported as tables, figures, and new datasets using the techniques that will be covered in this book.

1.2   R Objects

1.2.1   Vectors

When working with real data, it is necessary to keep track of sets of numbers that represent different types of measurements. Therefore, the fundamental type of object in R is a *vector*, which can contain multiple values. The single-value objects created in the previous section were actually vectors with a length of one.

As a more realistic example, consider data on forest canopy cover (the percent of the sky that is obscured by tree leaves, branches, and boles as seen from the ground) collected in the field. Data are usually collected from multiple field plots, and each plot may be revisited and resampled multiple times. We will start by creating a small dataset with 12 measurements of forest cover. The `c()` (combine) function is used to create a vector object named `cover` that contains these data. Any number of comma-separated data values can be provided as arguments to `c()`.

```
cover <- c(63, 86, 23, 77, 68, 91, 43, 76, 69, 12, 31, 78)
cover
##  [1] 63 86 23 77 68 91 43 76 69 12 31 78
```

An important element of coding in R (or any other programming language) is selecting names for the objects that are created. In R, object names must begin with a letter or a period, must contain only letters, numbers, underscores, and dots, and should not conflict with the names of R keywords and functions. For example, objects should never be named c because this name conflicts with the `c()` function.

It is up to the user to specify object names. Try to choose descriptive abbreviations that will help you remember the information contained in your R objects. However, variable names should not be so lengthy that they are cumbersome to read and type. In this case, `cover` is a simple and straightforward name to use, but `cc` or `cancov` or even `canopy_cover` would also be suitable depending on the programmer's preferences.

There are several widely used conventions for formatting object names, and it is best to select one and use it consistently. For example, consider an R object that will store topographic measurements of slope angle in radians. Some names based on commonly used case types include `slopeRad` (camel case), `slope_rad` (snake case), `slope-rad` (kebab case), and `SlopeRad` (Pascal case). Most of the code in this book is written in snake case because this is the style that the author learned when he started programming many years ago. The particular variable case that you choose matters less than applying it consistently and choosing good variable names.

Several functions can be used to create vectors containing regular sequences of numbers or repeated values. The seq() function generates a sequence of numbers from a beginning number to another number with values separated by a specified amount.

```
seq(from=1, to=12, by=1)
## [1]  1  2  3  4  5  6  7  8  9 10 11 12
seq(from=10, to=100, by=10)
## [1]  10  20  30  40  50  60  70  80  90 100
```

The colon operator (:) can also be used to generate integer sequences between two numbers.

```
1:10
## [1]  1  2  3  4  5  6  7  8  9 10
37:41
## [1] 37 38 39 40 41
```

The rep() function takes a vector and repeats the entire vector a specified number of times, or repeats each value in the vector.

```
rep(5, times=4)
## [1] 5 5 5 5
s1 <- 1:4
rep(s1, times=4)
## [1] 1 2 3 4 1 2 3 4 1 2 3 4 1 2 3 4
rep(s1, each=3)
## [1] 1 1 1 2 2 2 3 3 3 4 4 4
```

These examples provide a first look at how R *functions* work. One or more *arguments* are specified to determine the sequences that are generated, and the function outputs a vector. Because the function output is not assigned to an object, it is printed to the screen by default. We will discuss functions in more detail and look at some more complex examples later in the chapter.

Continuing with the canopy cover example, assume that there are four plots are coded 1, 2, 3, and 4, and each plot was measured in 2014, 2015, and 2016. The following code creates vectors containing the plot codes and years corresponding to each set of canopy cover observations.

```
plots <- 1:4
plot_codes <- rep(plots, times=3)
years <- rep(2014:2016, each=4)
plot_codes
## [1] 1 2 3 4 1 2 3 4 1 2 3 4
years
## [1] 2014 2014 2014 2014 2015 2015 2015 2015 2016 2016 2016
## [12] 2016
```

One of the most important features of R is that it uses vector arithmetic to carry out element-wise mathematical operations without having to write code for looping. If a mathematical operation is performed on two or more vectors of equal length, the operation will be carried out on the elements of each vector in sequence, returning a vector of the same length as the inputs. If one input is a single value and another is a longer vector, the single value will automatically be repeated for the entire length of the longer vector. Thus, the following two statements produce the same output.

```
cover + cover
## [1] 126 172  46 154 136 182  86 152 138  24  62 156
cover * 2
## [1] 126 172  46 154 136 182  86 152 138  24  62 156
```

Some additional examples of vector arithmetic are provided below. They add the same scalar to each canopy cover value, convert the canopy cover from percentage to proportion, and compute the mean of three canopy cover measurements for each combination of plot and year.

```
cover + 5
## [1] 68 91 28 82 73 96 48 81 74 17 36 83
cover / 100
## [1] 0.63 0.86 0.23 0.77 0.68 0.91 0.43 0.76 0.69 0.12 0.31
## [12] 0.78
cover2 <- c(59, 98, 28, 71, 62, 90, 48, 77, 74, 15, 38, 75)
cover3 <- c(91, 91, 33, 68, 59, 88, 44, 81, 72, 23, 44, 67)
tot_cover <- cover + cover2 + cover3
mean_cover <- tot_cover / 3
mean_cover
## [1] 71.00000 91.66667 28.00000 72.00000 63.00000 89.66667
## [7] 45.00000 78.00000 71.66667 16.66667 37.66667 73.33333
```

Vectors can also be supplied as arguments to functions to compute statistics such as the mean, sum, variance, and length. The examples below show how to use functions to calculate the sum, mean, and variance of canopy cover as well as the total number of values in the vector.

```
sum(cover)
## [1] 717
mean(cover)
## [1] 59.75
var(cover)
## [1] 678.3864
length(cover)
## [1] 12
```

The following examples show how vector operations can be used to calculate the mean and variance. The mean is the sum of the canopy cover values divided by the number of observations. The variance is the sum of square differences of each canopy cover and the mean canopy cover divided by the number of observations minus one.

```
meancover <- sum(cover) / length(cover)
meancover
## [1] 59.75
varcover <- sum((cover - meancover)^2) / (length(cover)-1)
varcover
## [1] 678.3864
```

Subsets of data can be selected by specifying one or more index numbers within square brackets. These index numbers are also referred to as subscripts. Positive numbers are used to include subsets in the output, and negative numbers are used to exclude subsets.

```
cover[1]
## [1] 63
cover[10]
## [1] 12
cover[c(1, 3, 8, 11)]
## [1] 63 23 76 31
cover[9:12]
## [1] 69 12 31 78
cover[-2]
##  [1] 63 23 77 68 91 43 76 69 12 31 78
```

Up to this point, we have been working with vectors that contain numerical values. However, vectors can also belong to different classes that contain other types of information. For example, logical vectors have only two possible values

(TRUE and FALSE) and belong to a different object class than the numeric vectors that we have been working with so far. Logical vectors can be used to select subsets that meet certain criteria. Here, a logical vector that indicates which observations have canopy cover values greater than 50% is generated. Then the logical vector is used to extract the subset of canopy cover observations greater than 50%. The class() function returns the object class to which the vector belongs.

```
class(cover)
## [1] "numeric"
cov_gt50 <- cover > 50
cov_gt50
##  [1]  TRUE  TRUE FALSE  TRUE  TRUE  TRUE FALSE  TRUE  TRUE
## [10] FALSE FALSE  TRUE
class(cov_gt50)
## [1] "logical"
cover[cov_gt50]
## [1] 63 86 77 68 91 76 69 78
```

The logical expression can also be nested inside the brackets to extract the subset using a single line of code.

```
high_cov <- cover[cover > 50]
high_cov
## [1] 63 86 77 68 91 76 69 78
```

In addition to numeric and logical vectors, vectors can also be composed of character data such as place names and plot codes.

```
plot_name <- c("Plot A4", "Plot A16", "Plot B2", "Plot B5", "Plot C11")
length(plot_name)
## [1] 5
class(plot_name)
## [1] "character"
```

There is also a special type of vector called a *factor* that is used for categorical data. To use the vector of plot names in an analysis comparing statistics across the different plots, it first needs to be converted into a factor. There are many more nuances to factors, but for now, it is sufficient to know that the factor object contains a levels attribute that specifies its categories.

```
plot_fact <- factor(plot_name)
class(plot_fact)
```

```
## [1] "factor"
plot_name
## [1] "Plot A4"  "Plot A16" "Plot B2"  "Plot B5"  "Plot C11"
plot_fact
## [1] Plot A4  Plot A16 Plot B2  Plot B5  Plot C11
## Levels: Plot A16 Plot A4 Plot B2 Plot B5 Plot C11
```

Numerical vectors can also be converted to factors. For example, there are some situations where plot numbers or years need to be treated as categories rather than continuous numerical measurements. The following code overwrites the plot_codes and year objects, effectively converting them from numeric vectors to factors.

```
plot_codes
##  [1] 1 2 3 4 1 2 3 4 1 2 3 4
years
##  [1] 2014 2014 2014 2014 2015 2015 2015 2015 2016 2016 2016
## [12] 2016
plot_codes <- factor(plot_codes)
years <- factor(years)
plot_codes
##  [1] 1 2 3 4 1 2 3 4 1 2 3 4
## Levels: 1 2 3 4
years
##  [1] 2014 2014 2014 2014 2015 2015 2015 2015 2016 2016 2016
## [12] 2016
## Levels: 2014 2015 2016
```

The NA symbol is a special value used in R to indicate missing data. When processing and managing data, it is critical that missing data be appropriately flagged as NA rather than treated as zero or some other arbitrary value. Most R functions have built-in techniques for handling missing data, and as we will see in later examples, the user must often choose the appropriate technique for a particular analysis.

In the example below, a vector is created containing four numerical values and two NA codes. The is.na() function returns a vector of logical values indicating whether or not each element in the vector is NA. Because logical values of TRUE and FALSE have equivalent numerical values of 1 and 0, summing the logical vector returns the total number of NA values.

```
myvector <- c(2, NA, 9, 2, 1, NA)
is.na(myvector)
## [1] FALSE  TRUE FALSE FALSE FALSE  TRUE
```

```
sum(is.na(myvector))
## [1] 2
```

By default, many functions will return NA if the input data contain NA values. When the na.rm argument is set to TRUE, the NA values are removed and the mean is computed using only the valid observations.

```
mean(myvector)
## [1] NA
mean(myvector, na.rm = TRUE)
## [1] 3.5
```

Dealing with missing data is an important aspect of most data science workflows. A common mistake is to assume that missing data are equivalent to a zero value. For example, rainfall observations may be missing from a meteorological dataset for several weeks because of an equipment malfunction or data loss. However, it should not be assumed that no rainfall occurred during these weeks just because no measurements were taken. Missing data should always be specified as NA in R, which will allow the analyst to identify them and take appropriate steps to account for them when summarizing and analyzing the data.

1.2.2 Matrices and lists

Multiple vectors of the same length can be combined to create a matrix, which is a two-dimensional object with columns and rows. All of the values in a matrix must be the same data type (e.g., numeric, character, or logical). The cbind() function combines the vectors as columns and the rbind() function combines the vectors as rows. There is also a matrix() function that reformats data from a vector into a matrix. The dim() function returns the number of rows and columns in the matrix.

```
mat1 <- cbind(cover, cover2, cover3)
mat1
##      cover cover2 cover3
## [1,]    63     59     91
## [2,]    86     98     91
## [3,]    23     28     33
## [4,]    77     71     68
## [5,]    68     62     59
## [6,]    91     90     88
## [7,]    43     48     44
## [8,]    76     77     81
```

```
## [9,]     69        74        72
## [10,]    12        15        23
## [11,]    31        38        44
## [12,]    78        75        67
class(mat1)
## [1] "matrix" "array"
dim(mat1)
## [1] 12  3
```

Subsets of matrix elements can be extracted by providing row and column indices in [row, column].

```
mat1[1,1]
## cover
##    63
```

Leaving one of these values blank will return all rows or columns, as shown in the examples below.

```
mat1[1:3,]
##        cover cover2 cover3
## [1,]     63     59     91
## [2,]     86     98     91
## [3,]     23     28     33
mat1[,2]
##  [1] 59 98 28 71 62 90 48 77 74 15 38 75
```

Lists are ordered collections of objects. They are created using the list() function and elements can be extracted by index number or component name. Lists differ from vectors and matrices in that they can contain a mixture of different data types. The example below creates a two-element list containing a vector and a matrix.

```
l1 <- list(myvector = cover, mymatrix = mat1)
l1
## $myvector
##  [1] 63 86 23 77 68 91 43 76 69 12 31 78
##
## $mymatrix
##        cover cover2 cover3
## [1,]     63     59     91
## [2,]     86     98     91
## [3,]     23     28     33
```

```
##  [4,]     77      71      68
##  [5,]     68      62      59
##  [6,]     91      90      88
##  [7,]     43      48      44
##  [8,]     76      77      81
##  [9,]     69      74      72
## [10,]     12      15      23
## [11,]     31      38      44
## [12,]     78      75      67
```

List elements can be extracted by providing either an element number or name inside of double brackets.

```
l1[[1]]
##  [1] 63 86 23 77 68 91 43 76 69 12 31 78
l1[["myvector"]]
##  [1] 63 86 23 77 68 91 43 76 69 12 31 78
```

List elements can also be extracted using the $ operator followed by the element name.

```
l1$myvector
##  [1] 63 86 23 77 68 91 43 76 69 12 31 78
```

1.2.3   Data frames

Data frames are like matrices in that they have a rectangular format consisting of columns and rows, and they are like lists in that they can contain columns with different data types such as numeric, logical, character, and factor. In general, the rows of a data frame represent observations (e.g., measurements taken at different locations and times), whereas the columns contain descriptive labels and variables for each observation.

The example below uses the data.frame() function to create a simple data frame consisting of three different canopy cover measurements taken at four plots for 3 years (12 observations total). The attributes() function returns the dimensions of the data frame along with the variable names.

```
cover_data <- data.frame(plot_codes, years, cover, cover2, cover3)
attributes(cover_data)
## $names
## [1] "plot_codes" "years"      "cover"      "cover2"
## [5] "cover3"
```

```
##
## $class
## [1] "data.frame"
##
## $row.names
##  [1]  1  2  3  4  5  6  7  8  9 10 11 12
```

The summary() function returns an appropriate statistical summary of each
variable in the data frame. For columns of factors such as plot_codes, the
counts of observations corresponding to each level are displayed. For columns
of numerical data such as cover, several summary statistics are provided to
describe the distribution of the values.

```
summary(cover_data)
##  plot_codes   years        cover           cover2
##  1:3          2014:4   Min.   :12.00   Min.   :15.00
##  2:3          2015:4   1st Qu.:40.00   1st Qu.:45.50
##  3:3          2016:4   Median :68.50   Median :66.50
##  4:3                   Mean   :59.75   Mean   :61.25
##                        3rd Qu.:77.25   3rd Qu.:75.50
##                        Max.   :91.00   Max.   :98.00
##      cover3
##  Min.   :23.00
##  1st Qu.:44.00
##  Median :67.50
##  Mean   :63.42
##  3rd Qu.:82.75
##  Max.   :91.00
```

The columns of a data frame can be accessed like list elements with the $
operator.

```
cover_data$plot_codes
##  [1] 1 2 3 4 1 2 3 4 1 2 3 4
## Levels: 1 2 3 4
cover_data$cover
##  [1] 63 86 23 77 68 91 43 76 69 12 31 78
```

Values in data frames can also be accessed using matrix-style indexing of rows
and columns.

```
cover_data[3, 4]
## [1] 28
```

```
cover_data[1:3,]
##    plot_codes years cover cover2 cover3
## 1           1  2014    63     59     91
## 2           2  2014    86     98     91
## 3           3  2014    23     28     33
cover_data[,4]
##  [1] 59 98 28 71 62 90 48 77 74 15 38 75
cover_data[,"cover2"]
##  [1] 59 98 28 71 62 90 48 77 74 15 38 75
```

Conditional statements can be used to query data records meeting certain conditions. Note that when referencing columns of a data frame, it is necessary to specify the data frame and reference the columns using the $ operator.

```
cover_data[cover_data$years == 2015,]
##    plot_codes years cover cover2 cover3
## 5           1  2015    68     62     59
## 6           2  2015    91     90     88
## 7           3  2015    43     48     44
## 8           4  2015    76     77     81
cover_data[cover_data$cover > 70,]
##     plot_codes years cover cover2 cover3
## 2            2  2014    86     98     91
## 4            4  2014    77     71     68
## 6            2  2015    91     90     88
## 8            4  2015    76     77     81
## 12           4  2016    78     75     67
```

A conditional statement based on the & (logical and) operator requires that both conditions are met.

```
cover_data[cover_data$cover > 70 & cover_data$cover3 > 70,]
##    plot_codes years cover cover2 cover3
## 2           2  2014    86     98     91
## 6           2  2015    91     90     88
## 8           4  2015    76     77     81
```

A conditional statement based on the | (logical or) operator requires that at least one of the conditions is met.

```
cover_data[cover_data$cover > 70 | cover_data$cover3 > 70,]
##    plot_codes years cover cover2 cover3
## 1           1  2014    63     59     91
```

```
## 2              2  2014    86    98    91
## 4              4  2014    77    71    68
## 6              2  2015    91    90    88
## 8              4  2015    76    77    81
## 9              1  2016    69    74    72
## 12             4  2016    78    75    67
```

1.3   R Functions

An R function takes one or more *arguments* as inputs and produces an object
as the output. Thus far, a variety of relatively simple functions have been
used to do basic data manipulation and print results to the console. Moving
forward, the key to understanding how to do more advanced geospatial data
processing, visualization, and analysis in R is learning, which functions to use
and how to specify the appropriate arguments for those functions. This section
will illustrate the use of some functions to generate basic graphics and carry
out standard statistical analyses.

1.3.1   Data input and graphics

Data are imported from an external file named `"camp32.csv,"` which contains
data in a comma-delimited format. The `read.csv()` function reads in the file
and outputs to a data frame, which is stored as an object called `camp32_data`.
The `file` argument indicates the name of the file to read. Make sure this
file is in your working directory so that it can be read without specifying
any additional information about its location. The most straightforward way
to do this is to create an R project associated with a specific folder in your
computer's file system. Then copy your input files to this folder and it will
become the default location for file input and output. See the section in the
Appendix on Managing RStudio Projects for more details.

```
camp32_data <- read.csv(file = "camp32.csv")
summary(camp32_data)
##       DATE            PLOT_ID              LAT
##  Length:36         Length:36         Min.   :48.84
##  Class :character  Class :character  1st Qu.:48.85
##  Mode  :character  Mode  :character  Median :48.85
##                                      Mean   :48.85
##                                      3rd Qu.:48.85
##                                      Max.   :48.86
```

```
##         LONG              DNBR              CBI
##  Min.    :-115.2   Min.    :-34.0   Min.    :0.780
##  1st Qu.:-115.2    1st Qu.: 85.5    1st Qu.:1.302
##  Median :-115.2    Median :298.0    Median :2.125
##  Mean    :-115.2   Mean    :333.2   Mean    :1.906
##  3rd Qu.:-115.2    3rd Qu.:561.0    3rd Qu.:2.485
##  Max.    :-115.2   Max.    :771.0   Max.    :2.700
##    PLOT_CODE
##  Length:36
##  Class :character
##  Mode  :character
##
##
##
```

These data include field and remote sensing measurements of wildfire severity for a set of plots on the Camp 32 fire in northwestern Montana. Higher values of the wildfire severity indices generally indicate that a higher proportion of trees were damaged or killed by the fire. Some of the plots were subjected to fuel treatments that removed overstory trees by mechanical thinning, reduced live and dead surface fuels with low-severity prescribed burns, or combined thinning and prescribed burning. Visualizing and analyzing these data can allow us to see whether these treatments moderated the effects of the wildfire. More information about the Camp32 fire and the data are available in Wimberly et al. (2009).

The following data columns will be used.

- CBI: Composite Burn Index, a field-measured index of fire severity
- DNBR: Difference Normalized Burn Ratio, a remotely sensed index of fire severity
- PLOT_CODE: Fuel treatments that were implemented before the wildfire, U = untreated, T = thinning, TB = thinning and prescribed burning

One question is whether the field-based and satellite-based metrics provide similar information about wildfire severity. We can explore this by using the plot() function to generate a scatterplot of CBI versus DNBR. Two vectors of data are specified as the x and y arguments. Note that the $ operator must be used to access the columns of the camp32_data data frame. The result shows that there is a relatively strong association between these two variables and suggests that the relationship may be non-linear (Figure 1.1).

```
plot(x = camp32_data$CBI, y = camp32_data$DNBR)
```

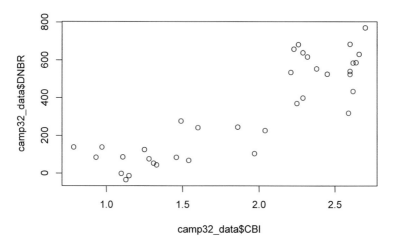

FIGURE 1.1
Scatterplot of the CBI and DNBR fire severity indices.

When a plotting function is run, its output is sent to a graphical device. In RStudio, the built-in graphical device is located in the lower right-hand pane in the *Plots* tab. The plot can be copied or saved to a graphics file using the *Export* button, and it can be expanded using the *Zoom* button.

To make a more interpretable plot, a title and axis labels can be specified as additional arguments to the `plot()` function (Figure 1.2).

```
plot(x = camp32_data$CBI, y = camp32_data$DNBR,
     main = "Camp 32 Fire Severity",
     xlab = "CBI",
     ylab = "DNBR")
```

At this initial data exploration stage, it is also helpful to explore the distributions of the variables of interest using histograms. The `hist()` function is used in this example to create a histogram of the DNBR values (Figure 1.3). Because a histogram is a univariate plot, only a single vector of data is provided as the x argument along with a title and x-axis label.

```
hist(x = camp32_data$DNBR,
     main = "Differenced Normalized Burn Ratio",
     xlab = "DNBR")
```

A boxplot can be used to visualize the distributions of fire severity in plots with different types of fuel treatments. The `boxplot()` function generates this plot, and the arguments used here are different from the preceding examples.

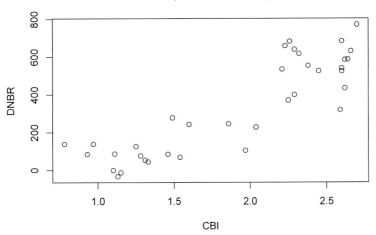

FIGURE 1.2
Scatterplot of the CBI and DNBR fire severity indices with axis labels and title.

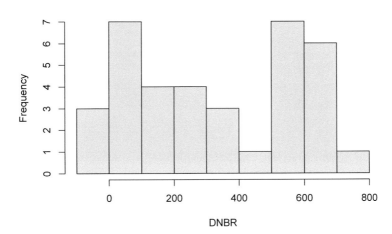

FIGURE 1.3
Histogram of the DNBR fire severity index.

The first argument is a formula, DNBR ~ PLOT_CODE, where the value to the left of the ~ (tilde) operator is the dependent variable and one or more values on the right-hand side are independent variables. The data argument specifies the data frame containing these columns, and additional arguments provide

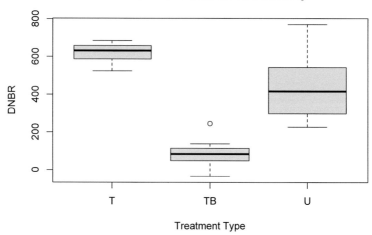

FIGURE 1.4
Boxplot showing the distribution of the DNBR fire severity index.

title and axis labels. Compared to the untreated plots, the thinned plots have slightly higher DNBR, whereas the thinned and burned plots have much lower DNBR (Figure 1.4).

```
boxplot(formula = DNBR ~ PLOT_CODE,
        data = camp32_data,
        main = "Treatment Effects on Fire Severity",
        xlab = "Treatment Type",
        ylab = "DNBR")
```

Knowing the arguments that can be specified for a specific function is very important. The quickest way to learn more about an R function is to display its documentation using the help() function. This function provides access to manual pages with information about function arguments, details about what the function does, and a description of the output that is produced. In RStudio, the help pages are displayed in the *Help* tab in the lower right-hand window.

```
help(plot)
help(hist)
help(boxplot)
```

1.3.2 Statistical analysis

Functions are also used to carry out statistical tests in R. For example, the
t.test() function can be used to compute confidence intervals for the mean
of a variable. The x argument is the vector of data to summarize. The output
of the function is an object of class htest that contains information about the
results of the statistical test. When the output of a function like t.test() is
not assigned to an object as in the following examples, the print() function
is invoked for the resulting htest object. The print() method for this object
class generates a summary of the statistical test results and outputs them to
the console.

```
t.test(x = camp32_data$DNBR)
##
##   One Sample t-test
##
## data:   camp32_data$DNBR
## t = 7.963, df = 35, p-value = 2.286e-09
## alternative hypothesis: true mean is not equal to 0
## 95 percent confidence interval:
##   248.2493 418.1396
## sample estimates:
## mean of x
##   333.1944
```

The conf.level argument can be used to specify the probability association
with the confidence intervals. Note that in the previous example, no conf.level
argument was provided, but a 95% confidence interval was produced.

```
t.test(x=camp32_data$DNBR, conf.level=0.9)
##
##   One Sample t-test
##
## data:   camp32_data$DNBR
## t = 7.963, df = 35, p-value = 2.286e-09
## alternative hypothesis: true mean is not equal to 0
## 90 percent confidence interval:
##   262.4982 403.8907
## sample estimates:
## mean of x
##   333.1944
```

How does the t.test() function know what confidence level to use if we do not
specify the conf.level argument? In many cases, arguments have a default
value. If the argument is not specified in the function call, then the default

value is used. When using statistical functions in R, it is particularly important to study the documentation to learn the default values and decide if they are appropriate for the analysis. The default values are specified at the beginning of the function's help page in the usage section.

```
help(t.test)
```

A two-sample t-test can be used to compare the mean DNBR values for treated versus untreated plots. The same function is used with two arguments, x and y, which are the two vectors of data to be compared. In the code below, two new vectors are created from the camp32_data data frame. U_DNBR contains DNBR values from the untreated plots and TB_DNBR contains DNBR values from the thinned and burned plots. Within the square brackets, the logical statement to the left of the comma specifies which rows to include, the character string to the right of the comma specifies which column to include.

```
U_DNBR <- camp32_data[camp32_data$PLOT_CODE == "U", "DNBR"]
TB_DNBR <- camp32_data[camp32_data$PLOT_CODE == "TB", "DNBR"]
t.test(x=U_DNBR, y=TB_DNBR)
##
##   Welch Two Sample t-test
##
## data:   U_DNBR and TB_DNBR
## t = 7.0902, df = 14.174, p-value = 5.058e-06
## alternative hypothesis: true difference in means is not equal to 0
## 95 percent confidence interval:
##   248.8065 464.2602
## sample estimates:
## mean of x mean of y
##   436.3333   79.8000
```

A linear regression analysis can be used to explore the relationship between the field-based burn severity index (CBI) and the satellite-derived burn severity index (DNBR). Ordinary least squares regression is implemented with the lm() function. This function requires at least two arguments, a formula and a data argument, and returns an object belonging to class lm. Note that lm objects, like the htest objects produced by t.test() are really list objects whose elements contain various pieces of information about the analysis results. The names of these elements can be obtained using the attributes() function.

```
camp32_lm <- lm(formula = CBI ~ DNBR,
                data = camp32_data)
class(camp32_lm)
## [1] "lm"
```

```
attributes(camp32_lm)
## $names
##  [1] "coefficients"  "residuals"      "effects"
##  [4] "rank"          "fitted.values" "assign"
##  [7] "qr"            "df.residual"   "xlevels"
## [10] "call"          "terms"         "model"
##
## $class
## [1] "lm"
```

To obtain more interpretable results, additional functions are needed to summarize the lm object. In this case, the print() method displays only the original function and the regression coefficients. There is also a generic summary() function.

```
camp32_lm
##
## Call:
## lm(formula = CBI ~ DNBR, data = camp32_data)
##
## Coefficients:
## (Intercept)          DNBR
##    1.178530      0.002183
```

For lm objects, the summary() method provides much more detail, including statistical tests for the coefficients and the overall model as well as the R^2 measure of model fit. Here, the R^2 value of 0.76 indicates a moderately strong association between these two indices.

```
summary(camp32_lm)
##
## Call:
## lm(formula = CBI ~ DNBR, data = camp32_data)
##
## Residuals:
##      Min       1Q   Median       3Q      Max
## -0.69976 -0.20237  0.00781  0.18838  0.71515
##
## Coefficients:
##              Estimate Std. Error t value Pr(>|t|)
## (Intercept) 1.1785297  0.0867154   13.59 2.66e-15 ***
## DNBR        0.0021828  0.0002089   10.45 3.75e-12 ***
## ---
```

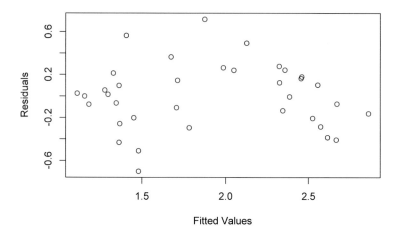

FIGURE 1.5
Scatterplot of residuals versus fitted values from linear regression of CBI versus DNBR.

```
## Signif. codes:
## 0 '***' 0.001 '**' 0.01 '*' 0.05 '.' 0.1 ' ' 1
##
## Residual standard error: 0.3103 on 34 degrees of freedom
## Multiple R-squared:  0.7625, Adjusted R-squared:  0.7555
## F-statistic: 109.2 on 1 and 34 DF,  p-value: 3.746e-12
```

The fitted values and residuals of the regression model can be extracted using the `fitted()` and `resid()` functions. The scatterplot of these variables is a common diagnostic of model fit (Figure 1.5). In this case, the arched shape of the point cloud indicates nonlinearity in the relationship between the two severity indices.

```
plot(x = fitted(camp32_lm), y = resid(camp32_lm),
     xlab = "Fitted Values",
     ylab = "Residuals")
```

1.4 Tips for Programming in R

The following block of code contains several lines that begin with a hash symbol (#) followed by descriptive text. Whenever the R interpreter encounters the # symbol, the rest of that line is ignored. Therefore, # is used to indicate comment lines that contain descriptive text rather than interpretable code. Whenever writing code in R or other computer languages, it is essential to include descriptive comments that explain what the code is intended to do and how it is expected to work. Comments are particularly important if you intend to revisit and reuse or code at some point in the future, or share it with others to use and modify. Most of the code presented in this book does not include comments. This is intended to save space and increase readability, as explanations of the code are already provided in the accompanying text. In later chapters, there will be some examples of how comments can be used to organize and document R scripts that implement longer and more complicated analytical workflows.

```r
###########################################################
# Example R script demonstrating the use of comments
###########################################################

# Read in fire severity data comma-delimited file
camp32_data <- read.csv(file = "camp32.csv")
# Box plot of DNBR distributions for different treatments
boxplot(formula = DNBR ~ PLOT_CODE,
        data = camp32_data,
        main = "Treatment Effects on Fire Severity",
        xlab = "Treatment Type",
        ylab = "DNBR")
# Extract vectors of untreated and thinned/burned DNBR
U_DNBR <- camp32_data[camp32_data$PLOT_CODE == "U", "DNBR"]
TB_DNBR <- camp32_data[camp32_data$PLOT_CODE == "TB", "DNBR"]
# t-test with null hypothesis of equal means
t.test(x=U_DNBR, y=TB_DNBR)
# Linear regression of CBI versus DNBR
camp32_lm <- lm(formula = CBI ~ DNBR,
                data = camp32_data)
summary(camp32_lm)
# Residuals versus fitted values diagnostic plot
plot(x = fitted(camp32_lm), y = resid(camp32_lm),
     xlab = "Fitted Values",
     ylab = "Residuals")
```

1.5   Practice

1. Create two new vectors, one containing all the odd numbers between 1 and 19, and the other containing all even numbers between 2 and 20. Add the two vectors element-wise to produce a single vector and then create a new vector that contains only numbers greater than 15. Finally, compute the mean value of this vector. If you do all the calculations right, the mean will equal 29.

2. Starting with the cover_data data frame, create a new data frame containing only the rows of data for plot number 3 and print the results to the screen.

3. From the Camp 32 dataset, extract a vector containing the CBI values from the untreated plots (treatment code U) and another containing the CBI values for the thinned/unburned plots (treatment code T). Conduct a two-sample t-test comparing the mean CBI for these two treatment types.

4. Generate a histogram plot of CBI values and compare it with the histogram of DNBR values.

5. Generate a new scatterplot based on the observed values of CBI on the x-axis and the predicted values of CBI from the linear regression on the y-axis. Make the plot have an aspect ratio of 1 (equal scales of x and y axes). To do this, you will need to specify an additional argument. Consult the help page for the base `plot()` function to find the name of the argument to use.

2

Graphics with ggplot2

This chapter presents the fundamentals of data visualization in R using the **ggplot2** package. One of R's strengths is that it empowers a large global community of developers to create and disseminate new functionality through user-contributed packages. As a result, there are many packages that improve on the data manipulation and plotting capabilities included with the base R installation. Each package provides a set of functions with accompanying documentation and datasets. In this chapter, we will begin to explore several packages that are part of the **tidyverse** collection of data science tools. For simplicity, we will start by working with non-spatial datasets. In upcoming chapters, these techniques will be extended to generate maps and analyze geospatial data.

The `library()` function is used to load R packages. If these packages are not yet on your computer, you will need to install them using the `install.packages()` function or the installation tools available in RStudio under Tools > Install Packages.

```
library(ggplot2)
library(dplyr)
library(readr)
library(readxl)
```

Nearly every script requires loading one or more packages. Although a package only needs to be installed once, it must be loaded with `library()` every time a new R session is started. Therefore, it is good practice to include the necessary code at the beginning of the script file. This approach ensures that the packages are loaded at the beginning of your session and makes it easier to see which packages are being used in the script.

The data used in this chapter are meteorological observations from the Oklahoma Mesonet, a network of environmental monitoring stations distributed throughout Oklahoma (`https://weather.ok.gov/`). The data are provided in a comma-separated values (CSV) file, which can be read and stored as a data frame object using the `read_csv()` function from the **readr** package. This function does essentially the same thing as the base `read.csv()` function that was used in the last tutorial, but has a few added features and is faster for reading large csv files.

```
mesosm <- read_csv("mesodata_small.csv", show_col_types = FALSE)
class(mesosm)
## [1] "spec_tbl_df" "tbl_df"        "tbl"           "data.frame"
mesosm
## # A tibble: 240 x 9
##      MONTH   YEAR STID    TMAX   TMIN  HMAX   HMIN   RAIN
##      <dbl> <dbl> <chr> <dbl> <dbl> <dbl> <dbl> <dbl>
## 1        1   2014 HOOK    49.5   17.9  83.0   29.0   0.17
## 2        2   2014 HOOK    47.2   17.1  88.3   39.9   0.3
## 3        3   2014 HOOK    60.7   26.1  79.0   25.4   0.31
## 4        4   2014 HOOK    72.4   39.3  81.8   21.0   0.4
## 5        5   2014 HOOK    84.4   48.3  75.4   18.8   1.25
## 6        6   2014 HOOK    90.7   61.9  90.9   28.8   3.18
## 7        7   2014 HOOK    90.6   64.7  88.2   33.1   2.58
## 8        8   2014 HOOK    95.8   64.5  85.4   21.8   0.95
## 9        9   2014 HOOK    84.3   58.0  91.2   36.4   1.48
## 10      10   2014 HOOK    76.1   44.9  85.7   28.1   1.72
## # ... with 230 more rows, and 1 more variable: DATE <date>
```

Note that the `mesosm` data frame has multiple classes including `tbl` and `tbl_df`. It is an enhanced version of a data frame called a *tibble* that is part of the *tidyverse*, a collection of R packages for data science. A tibble is identical to a data frame but includes some additional features. For example, the default print method for a tbl object provides an abbreviated view of the first few rows of the data frame rather than trying to print all the data to the screen. In most cases, tbl objects can be used in exactly the same way as data frame objects, and these object types will be treated as synonymous throughout the book.

There are functions in R for importing data from just about any external file type. For example, the `read_excel()` function from the **readxl** package can be used to import data from XLS and XLSX files. Note that these spreadsheet files can contain multiple sheets, so it may be necessary to specify the sheet containing the data. In this example, the first sheet contains a data dictionary and the second sheet contains the data. Additional arguments can also be provided to extract data from a specific range of cells.

```
mesosm2 <- read_excel("mesodata_small.xlsx", sheet=2)
class(mesosm2)
## [1] "tbl_df"        "tbl"           "data.frame"
mesosm2
```

```
## # A tibble: 240 x 9
##    MONTH  YEAR STID   TMAX  TMIN  HMAX  HMIN  RAIN
##    <dbl> <dbl> <chr> <dbl> <dbl> <dbl> <dbl> <dbl>
##  1     1  2014 HOOK   49.5  17.9  83.0  29.0  0.17
##  2     2  2014 HOOK   47.2  17.1  88.3  39.9  0.3
##  3     3  2014 HOOK   60.7  26.1  79.0  25.4  0.31
##  4     4  2014 HOOK   72.4  39.3  81.8  21.0  0.4
##  5     5  2014 HOOK   84.4  48.3  75.4  18.8  1.25
##  6     6  2014 HOOK   90.7  61.9  90.9  28.8  3.18
##  7     7  2014 HOOK   90.6  64.7  88.2  33.1  2.58
##  8     8  2014 HOOK   95.8  64.5  85.4  21.8  0.95
##  9     9  2014 HOOK   84.3  58.0  91.2  36.4  1.48
## 10    10  2014 HOOK   76.1  44.9  85.7  28.1  1.72
## # ... with 230 more rows, and 1 more variable: DATE <dttm>
```

The data contain monthly summaries of several meteorological variables from 2014–2018. We will work with the following data columns.

- STID: Station ID code
- TMAX: Monthly mean of maximum daily temperature (°F)
- TMIN: Monthly mean of minimum daily temperature (°F)
- RAIN: Cumulative monthly rainfall (inches)
- DATE: Date of observation

There are a couple of new classes in these imported data frames. In the mesosm data frame, the DATE column belongs to the date class. In the mesosm2 data frame, the DATE column belongs to the dttm (date/time) class. Don't worry about these details for now—the functions in **ggplot2** will know how to handle these classes automatically. An upcoming chapter will provide more information about how to import and manipulate date objects.

2.1 Creating a Simple Plot

One of the most common types of scientific graphics is a time series plot, in which the date or time element is on the x-axis, and the measured variable of interest is on the y-axis. The data values are usually connected by a line to indicate progression through time.

The following code uses the filter() function from the **dplyr** package to extract the rows containing meteorological data for the Mount Herman station (MTHE). More details about filter() and other **dplyr** functions will be provided in Chapter 4. Here, the output of filter() is assigned to a new object called mesomthe. Then, a time series graph of monthly rainfall is generated

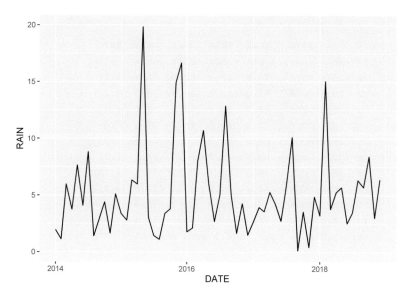

FIGURE 2.1
Line graph of monthly rainfall for Mount Herman.

(Figure 2.1). By default, the graphical output is written to the Plot tab in the RStudio GUI.

```
mesomthe <- filter(mesosm, STID =="MTHE")
ggplot(data = mesomthe) +
  geom_line(mapping = aes(x = DATE, y = RAIN))
```

The `ggplot()` function creates a coordinate system onto which data can be plotted. The first argument to `ggplot()` is `data`, the dataset to plot. Running only the first line, `ggplot(data = mesomthe)`, would create just the blank coordinate system. To add data and modify the appearance of the graph, an additional function from the **ggplot2** package is used. In the previous example `geom_line()` adds lines to the plot. The `+` symbol indicates that the line will be added to the coordinate system created by `ggplot()` using the `mesomthe` data frame

The `mapping` argument specifies the aesthetic mapping to use. Aesthetic mappings indicate which columns in the dataset get used for (or "mapped to") various features of the plot. The mapping is always specified by the `aes()` function. In this example, the code specifies that the `DATE` column contains x-axis values and the `RAIN` column contains y-axis values.

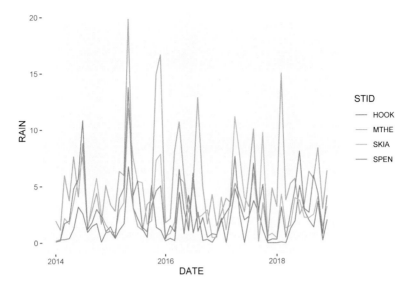

FIGURE 2.2
Line graph of monthly rainfall with a different line color for each station.

2.2 Aesthetic Mappings

In the previous example, data columns were mapped to the x and y axes. To visualize a third column of data, it needs to be mapped to some other aspect of the plot. To illustrate, we will use the full mesosm dataset, which contains five years of monthly meteorological data from four sites: HOOK (Hooker in western OK), MTHE (Mount Herman in southeastern OK), SKIA (Skiatook in northeastern OK, and SPEN (Spencer in central OK).

Because there are four sites in this dataset, there needs to be a way to distinguish lines for the sites. A common choice is to show different lines with different colors. Here the STID column is mapped to the color aesthetic (Figure 2.2).

```
ggplot(data = mesosm) +
  geom_line(mapping = aes(x = DATE,
                          y = RAIN,
                          color = STID))
```

Taking a critical look at 2.2, it is possible to differentiate the colored lines. However, they overlap considerably, and the time series of the different stations

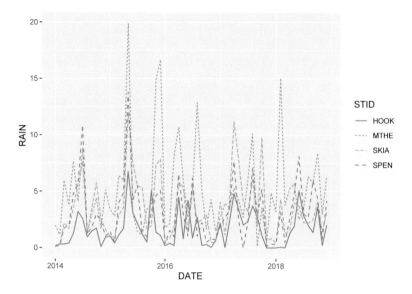

FIGURE 2.3
Line graph of monthly rainfall with a different line color and line pattern for
each station.

are difficult to identify and compare. We can experiment with different aesthetics to see if the graph can be improved. One idea is to try mapping multiple aesthetics to the same column. In this example, each site has a different line pattern as well as a different color (Figure 2.3).

```
ggplot(data = mesosm) +
  geom_line(mapping = aes(x = DATE,
                          y = RAIN,
                          color = STID,
                          linetype = STID))
```

Mapping aesthetics based on patterns in addition to (or instead of) color is important in many situations. For example, pattern-based aesthetics can be interpreted by color-blind individuals and reproduced in black and white. However, in this situation, they don't really help with distinguishing the overlapping lines in the graph.

Other aspects of the graph's aesthetics can be manipulated. An aesthetic can be set to a fixed value by defining that aesthetic outside of the aes() function. The following code increases the width of every line slightly to make them easier to see. In this example, the default line size of 0.5 is increased to 1.0

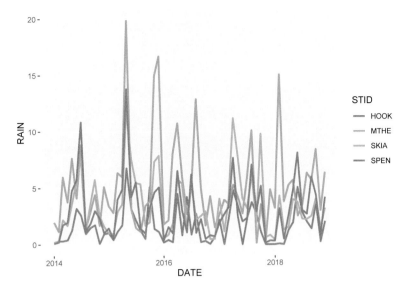

FIGURE 2.4
Line graph of monthly rainfall with a different line color for each station and
increased line size.

(Figure 2.4). The colors of the thicker lines are a bit easier to distinguish from
one another.

```
ggplot(data = mesosm) +
  geom_line(mapping = aes(x = DATE,
                          y = RAIN,
                          color = STID),
            size = 1.0)
```

2.3   Facets

In this sample dataset, the rainfall values from all the stations fall within the
same range of values. No matter how the plot aesthetics are modified, they will
be crowded and difficult to view on a single set of plot axes. An alternative is
to organize these data into multiple plots using *facets*. Faceting splits the data
into subsets based on one or more columns in the data frame and creates a
separate chart for each subset.

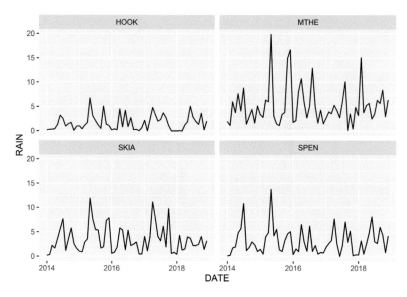

FIGURE 2.5
Line graph of monthly rainfall with each station plotted in a different facet.

To facet by a single variable, the `facet_wrap()` function is used. Another +
operator is used at the end of the `geom_line()` function to indicate that all
of these functions are combined to generate the plot. In this example, the
`facets` argument takes the name of a single column. The `vars()` function is also
needed to convert the column name into a format recognized by `facet_wrap()`.
The `STID` column contains a character vector with four different station codes,
so four facets are generated (Figure 2.5).

```
ggplot(data = mesosm) +
  geom_line(mapping = aes(x = DATE, y = RAIN)) +
  facet_wrap(facets = vars(STID))
```

Because these are time series data, it is often helpful to arrange the subplots on
top of each other instead of side-by-side. This format makes it easy to compare
values from the different stations at a given time point by scanning vertically
across the subplots (Figure 2.6). The layout can be changed by using the `ncol`
or `nrow` argument to specify the number of columns or rows.

```
ggplot(data = mesosm) +
  geom_line(mapping = aes(x = DATE, y = RAIN)) +
  facet_wrap(facets = vars(STID), ncol = 1)
```

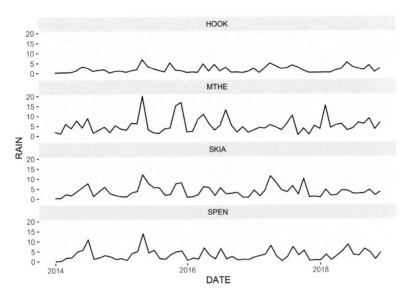

FIGURE 2.6
Line graph of monthly rainfall with facets arranged in a single column.

By default, the `facet_wrap()` function uses the same data range for each of the subplots. This approach clearly shows differences in the overall magnitude of rainfall across the subplots. In particular, rainfall at the Hooker Mesonet station in western Oklahoma is considerably lower than at the other stations. The rainfall values at Hooker are compressed within a relatively small portion of the y-axis, which makes it difficult to discern variation through time. Setting the `scales` argument to `"free_y"` allows the data in each facet to take up the entire vertical space. Alternate values for `scales` include `"free_x"` for scales to vary freely in the x dimension and `"free"` to allow them to vary freely in both dimensions. The resulting plot more clearly shows the relative variations in precipitation across the four stations (Figure 2.7).

```
ggplot(data = mesosm) +
  geom_line(mapping = aes(x = DATE, y = RAIN)) +
  facet_wrap(facets = vars(STID),
             ncol = 1,
             scales = "free_y")
```

One problem with allowing the scales to vary across facets is that readers may not carefully examine the y-axis of each plot. A naive viewer might look at Figure 2.7 and assume that rainfall is similar at all four locations based on the aesthetics alone. This approach should therefore be used with caution, and

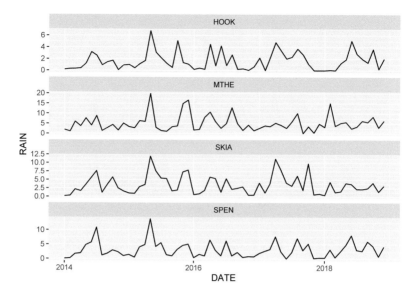

FIGURE 2.7
Line graph of monthly rainfall with free scales on the y-axis. Note that each
subplot has a different range of rainfall values.

the differences in axis ranges should be explicitly noted in the figure caption
and other accompanying text.

2.4 Geometric Objects

A `geom` is the geometrical object that a plot uses to represent data. There are
often multiple ways to represent the same data visually. For example, we have
been using the `geom_line()` function to visualize our time series of weather
data as line geometries. Line graphs can be very useful for mapping time series
such as weather data because the lines that connect the data points highlight
the changes that occur between time periods. Another common technique for
graphing data is to use a point for each combination of x and y values with
no connecting lines. Replacing the `geom_line` with the `geom_point()` function
generates the following graph (Figure 2.8).

```
ggplot(data = mesosm) +
  geom_point(mapping = aes(x = DATE, y = RAIN)) +
  facet_wrap(facets = vars(STID), ncol = 1)
```

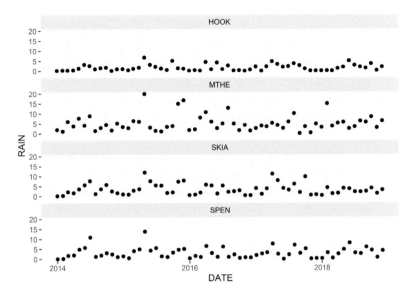

FIGURE 2.8
Point graph of monthly rainfall with facets arranged in a single column.

In most cases, using points alone is not a great choice for time series graphs because the connections between subsequent time periods are not as apparent as in the line graph. Conversely, lines should not be used with data that do not have a natural ordering because they will falsely imply connections between adjacent values in the graph. In some situations, it makes sense to plot data using multiple geometries. For example, with line graphs, it can be difficult to identify values at specific points in time-based only on changes in line direction.

Including both lines and points can be effective when it is important to emphasize the sequential nature of time series data and clearly see the individual measurements at each time interval (Figure 2.9). However, the graph is also more crowded and complicated than the simple line graph, and the additional point symbols may become a distraction if they are not really needed. In general, the key to effective scientific graphics is to include just enough details to effectively communicate the important patterns in the data while resisting the urge to add unnecessary embellishments.

```
ggplot(data = mesosm) +
  geom_line(mapping = aes(x = DATE, y = RAIN)) +
  geom_point(mapping = aes(x = DATE, y = RAIN)) +
  facet_wrap(facets = vars(STID),
             ncol = 1,
             scales = "free_y")
```

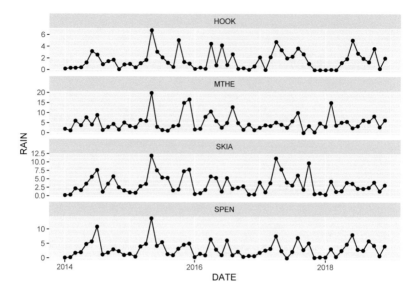

FIGURE 2.9
Combined line and point graph of monthly rainfall with facets arranged in a single column.

Note that some aesthetics can only be used with certain geoms. For example, points can have a `shape` aesthetic, but lines cannot. Conversely, lines can have a `linetype` aesthetic, but points cannot.

2.5   Scales

Scales control how data values are translated to visual properties. The default scale setting can be overridden to adjust details like axis labels and legend keys or to use a completely different translation from data to aesthetic.

The `labs()` function can be used to change the axis, legend, and plot labels. Additional arguments to `labs()` include `subtitle`, `caption`, `tag`, and any other aesthetics that have been mapped such as `color` or `linetype`. This example shows precipitation data from the four stations on a single plot with labels that specify the x-axis values, y-axis values, title for the legend of station colors, and an overall plot title (Figure 2.10).

```
ggplot(data = mesosm) +
  geom_line(mapping = aes(x = DATE,
```

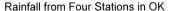
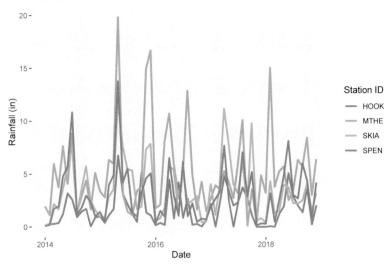

FIGURE 2.10
Line graph of monthly rainfall with a different line color for each station and modified labels.

```
                        y = RAIN,
                        color = STID),
            size = 1.0) +
  labs(x = "Date",
       y = "Rainfall (in)",
       color = "Station ID",
       title = "Rainfall from Four Stations in OK")
```

Including a graph title is not necessary in many cases. The title can usually be added later in a presentation or document file, and descriptive information can be provided in a separate caption. However, it is good practice to always add descriptive axis and legend titles with measurement units where appropriate. By default, these labels are just the column names from the data frame. Although the analyst may be familiar with these codes, they will usually not be interpretable by other people viewing the graph.

Tick marks and associated text labels indicate how the data are scaled along the axes of the graph. The ggplot() function uses an algorithm to optimize the default number and placement of the ticks along each axis. In general, there should be enough ticks to make it easy to associate plot aesthetics with axis values but not so many that the axis becomes cluttered and difficult to

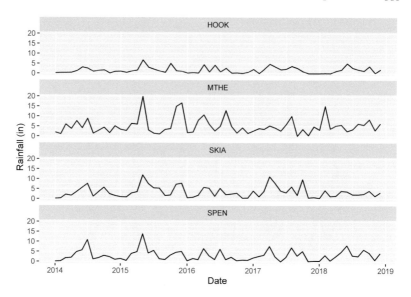

FIGURE 2.11
Line graph of monthly rainfall with modified tick marks and labels on the
x-axis.

interpret. Ticks are associated with round numbers to make the labels easier
to read.

In some situations, it is necessary to change the placement of the tick marks
and their labels. In the time series graphs produced so far, there is one tick
mark every two years on the x-axis, with each tick located on the first day of
the year. The wide spacing between ticks makes it difficult to determine what
year we are looking at when examining the precipitation patterns. It would be
helpful to change the x-axis so that there are ticks and labels for every year.

The `scale_x_date()` function allows modification of the tick marks and labels
on the x-axis. There are also `scale_x_continuous()` and `scale_x_discrete()`
functions that are appropriate when there are continuous or discrete variables
mapped to the axis, as well as versions of all these functions for the y-axis. Two
arguments are provided to `scale_x_date()`. The `breaks` argument indicates
where tick marks and labels will be placed, and the `date_label` argument
indicates how the dates will be formatted. The `%Y` code indicates that only the
year will be displayed. The resulting graph now shows one tick mark and label
per year (Figure 2.11).

```
datebreaks <- c("2014-01-01", "2015-01-01", "2016-01-01",
                "2017-01-01", "2018-01-01", "2019-01-01")
```

```
datebreaks <- as.Date((datebreaks))
ggplot(data = mesosm) +
  geom_line(mapping = aes(x = DATE, y = RAIN)) +
  facet_wrap(facets = vars(STID), ncol = 1) +
  labs(x = "Date",
       y = "Rainfall (in)",
       color = "Station ID") +
  scale_x_date(breaks = datebreaks, date_label = "%Y")
```

Remember that the DATE variable plotted on the x-axis is a Date object. This is why it is necessary to use the scale_x_date() function, and why the datebreaks vector needs to be converted from a character vector to a date vector using the as.Date() function. Chapter 4 will provide more information about date variables and how to manipulate them.

Scale functions are frequently used to specify the colors and symbols that will be mapped to the data values. For example, the scale_color_manual() function can be used to manually specify the colors for each value of the STID variable associated with the color aesthetic (Figure 2.12). Note that other aesthetics have their own scale functions, such as scale_linetype() and scale_size(). Additional examples of scale functions will be provided as more complex plots and maps are developed throughout the book.

```
ggplot(data = mesosm) +
  geom_line(mapping = aes(x = DATE,
                          y = RAIN,
                          color = STID),
            size = 1.0) +
  labs(x = "Date",
       y = "Rainfall (in)",
       color = "Station ID") +
  scale_x_date(breaks = datebreaks, date_label = "%Y") +
  scale_color_manual(values = c("firebrick",
                                "steelblue",
                                "olivedrab",
                                "gold"))
```

Colors can be specified in several ways in R. The previous example uses a character string with the name of each color name. A list of the 657 color names available in R can be obtained by running the function colors(). Colors can also be specified directly in terms of their red, green, and blue (RGB) components with a hexadecimal string of the form "#RRGGBB". With so many possibilities, selecting effective colors for scientific graphics is a formidable challenge. There are several resources available online that

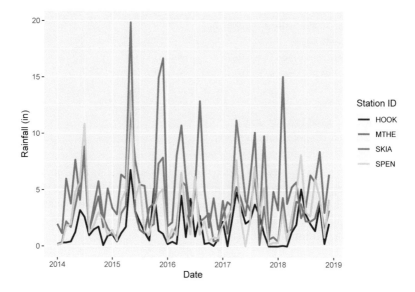

FIGURE 2.12
Line graph of monthly rainfall with manually selected line colors for each station.

list and display all of the named R colors, including the "R color cheatsheet" provided by the National Center for Ecological Analysis and Synthesis at `https://www.nceas.ucsb.edu/r-spatial-guides`. There are also R packages that can be used to automatically generate color palettes for graphs and maps. We will use several of these packages in upcoming chapters.

2.6 Themes

The default "look" of graphs created with `ggplot()` has some distinctive and recognizable features. The plot background is light gray with white gridlines that align with the axis ticks. There are no axis lines and the tick marks extend outward from the edge of the background grid. This default theme is relatively simple, and many types of graphs are easy to read against this background. However, users may prefer an alternative graph design or need to modify the plot elements to meet publication requirements. One way to change the theme is by specifying an alternative theme function. For example, the `theme_bw()` function uses a black grid on a white background (Figure 2.13).

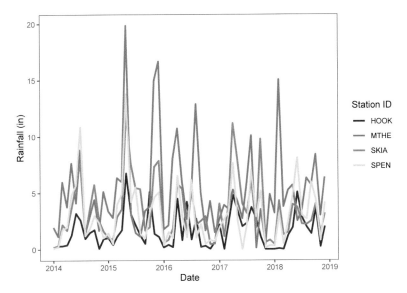

FIGURE 2.13
Line graph of monthly rainfall with a black-and-white theme.

```
ggplot(data = mesosm) +
  geom_line(mapping = aes(x = DATE,
                          y = RAIN,
                          color = STID),
            size = 1.0) +
  labs(x = "Date",
       y = "Rainfall (in)",
       color = "Station ID") +
  scale_color_manual(values = c("firebrick",
                                "steelblue",
                                "olivedrab",
                                "gold")) +
  scale_x_date(breaks = datebreaks, date_label = "%Y") +
  theme_bw()
```

The theme() function allows detailed formatting of plot components, including text, lines, and plot area. The following example shows how to modify various text elements including axis text, legend text, and the main title. The element_text() function is used to provide formatting details to each argument of the theme() function. Arguments to element_text() control text color, size, and angle. The hjust and vjust arguments control horizontal text justification along the x-axis and vertical text justification along the y-axis (0 = left, 0.5 = center, 1 = right). The face argument changes the font type from the

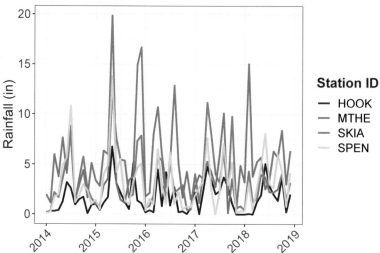

FIGURE 2.14

Line graph of monthly rainfall with modifications to text elements.

default of `plain` to `bold`, `italic`, or `bold.italic`. Compared to Figure 2.13, the resulting plot has slightly larger text (Figure 2.14). The plot, axis, and legend titles are dark blue and bolded. The plot title is also left justified. The x-axis title is suppressed by specifying `element_blank()`. The x-axis tick label text is angled to avoid crowding the larger text and justified to increase separation from the axis.

```
ggplot(data = mesosm) +
  geom_line(mapping = aes(x = DATE,
                          y = RAIN,
                          color = STID),
            size = 1.0) +
  labs(x = "Date",
       y = "Rainfall (in)",
       color = "Station ID",
       title = "Rainfall from Four Stations in OK") +
  scale_color_manual(values = c("firebrick",
                                "steelblue",
                                "olivedrab",
                                "gold")) +
  scale_x_date(breaks = datebreaks, date_label = "%Y") +
  theme_bw() +
  theme(axis.text.x = element_text(angle = 45,
```

```
                                size = 14,
                                hjust = 1,
                                vjust = 1),
        axis.text.y = element_text(size = 14),
        axis.title.x = element_blank(),
        axis.title.y = element_text(color = "darkblue",
                                    size = 16),
        legend.text = element_text(size = 14),
        legend.title = element_text(color = "darkblue",
                                    size = 16,
                                    face = "bold"),
        plot.title = element_text(color = "darkblue",
                                  size = 18,
                                  hjust = 0.5,
                                  face = "bold"),
)
```

A common problem with plots and maps is text that is too small to be easily readable. Always check the size of your text and consider the size and manner in which your plot will be displayed. Will it be embedded in a PDF document? Displayed on a website that will be viewed on a computer screen? Projected on a large screen in front of a lecture hall? Different text sizes may be required for each of these examples. You can use arguments to `theme()` like in the previous example to control text size and appearance in different sections of the graph. There are many more arguments to the `theme()` function that can be used to control other aspects of plot appearance, and additional examples will be provided in later chapters.

2.7 Combining ggplot Functions

Building a scientific graphic with ggplot involves a number of steps that are implemented with different types of functions.

1. Create a ggplot using `ggplot()`.
2. Add geometric representations of data to a plot using geoms.
3. Map data columns to plot aesthetics.
4. Split your dataset into subplots using facets.
5. Control the visual properties of your plot using scales.
6. Control other aspects of plot appearance by modifying themes.

Throughout the rest of the book, various combinations of these steps will be used to create charts and maps. By learning this process, you will be able to generate a wide variety of graphs and maps using diverse datasets. The upcoming chapters will provide many more examples of how to use `ggplot()` for data visualization. In addition to covering the technical aspects of coding, they will continue to explore how to design scientific graphics so that they summarize data effectively and communicate the resulting information clearly and accurately to viewers.

2.8   Other Types of Plots

The following subsections provide examples of how to generate other types of standard plots with `ggplot()`, including scatterplots, bar charts, histograms, and boxplots. Each example also uses a different theme function to show some of the options that are available in the **ggplot2** package.

2.8.1   Scatterplots

Scatterplots are used to show the relationship between two variables. They are similar to time series plots, but they use points for the geometry, and both axes are measured variables rather than just the y-axis. This example shows the relationship between monthly summaries of daily minimum and maximum temperatures (Figure 2.15). There is one point for each monthly record at each of the four meteorological stations. The `scale_color_manual()` function is used to assign a different color to each of the four stations. The plot shows that there is a very strong linear relationship between the monthly minimum and maximum temperatures at each station. The temperatures at Hooker are much higher than those at the other three stations.

```
ggplot(data = mesosm) +
  geom_point(mapping = aes(x = TMIN,
                           y = TMAX,
                           color = STID)) +
  labs(x = "Minimum Temperature (\u00B0F)",
       y = "Maximum Temperature (\u00B0F)",
       color = "Station ID") +
  scale_color_manual(values = c("firebrick",
                                "steelblue",
                                "olivedrab",
                                "gold")) +
  theme_minimal()
```

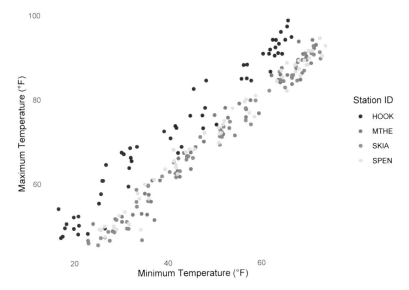

FIGURE 2.15
Scatterplot of monthly minimum and maximum temperatures at four weather stations.

2.8.2 Bar charts

Bar charts represent data values using the heights of horizontal or vertical bars, and the color and arrangement of the bars can be used to represent groupings within the data. The following example creates a simple bar chart to display the precipitation times series as bars rather than lines or dots (Figure 2.16). Other examples of grouped bar charts will be provided later in later chapters. Because the monthly precipitation values represent cumulative sums for each month, the relative heights of the bars provide an intuitive representation of the month-to-month variation.

```
ggplot(data = mesosm) +
  geom_col(mapping = aes(x = DATE, y = RAIN)) +
  labs(x = "Date",
       y = "Precipitation (in)") +
  facet_wrap(facets = vars(STID)) +
  theme_bw()
```

2.8.3 Histograms

A histogram is a graphical representation of the distribution of numerical data. The heights of the bars are proportional to the frequencies of observations within

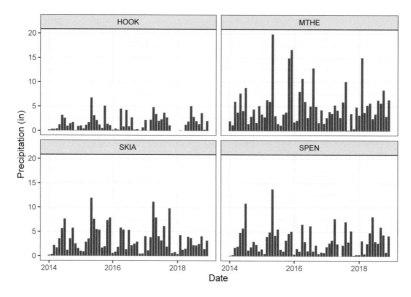

FIGURE 2.16
Bar charts of monthly rainfall at four weather stations.

different value ranges. The example below plots the frequency distribution
of rainfall at the Mount Herman station (Figure 2.17). Custom labels are
added to the axes, and the text is bolded and increased in size to make it
more readable. The graphs show that most months in the dataset have rainfall
between about 2–7 inches, but there are a few outlying months with rainfall
between 10 and 20 inches.

```
ggplot(data = mesomthe) +
  geom_histogram(aes(x = RAIN), bins = 10) +
  labs(x = "Precipitation (in)",
       y = "Count of observations") +
  theme_linedraw() +
  theme(axis.text.x = element_text(size = 12, face = "bold"),
        axis.text.y = element_text(size = 12, face = "bold"),
        axis.title.x = element_text(size = 14, face = "bold"),
        axis.title.y = element_text(size = 14, face = "bold"))
```

2.8.4   Boxplots

Like histograms, boxplots also display the distributions of data. Boxplots
provide a more simplified representation of the distribution than histograms,
and it is often easier to compare multiple variables using boxplots. This example

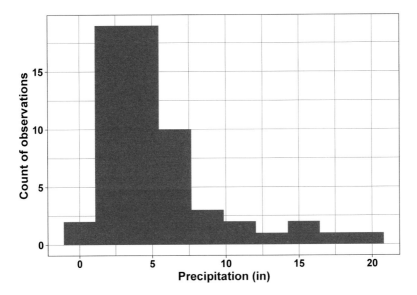

FIGURE 2.17
Histogram of monthly rainfall at Mt. Herman.

shows the rainfall distributions for the four mesonet stations (Figure 2.18). For boxplots generated with ggplot(), the horizontal line represents the median value, and the box represents the inter-quartile range (the 25th through 75th percentiles). The upper and lower whiskers extend to the largest and smallest value no further than 1.5 times the hinges (the edges of the box). Data with higher or lower values than the whiskers are called "outliers" and are plotted individually.

```
ggplot(data = mesosm) +
  geom_boxplot(aes(x = STID, y = RAIN)) +
  labs(x = "Station Code",
       y = "Precipitation (in)") +
  theme_classic()
```

Although this chapter has covered a lot of information, it has only begun to scratch the surface of what is possible with the **ggplot2** package (Wickham, 2016). For a complete listing of **ggplot2** functions, you can check out the online reference at https://ggplot2.tidyverse.org. Another helpful reference is the ggplot2 "cheatsheet", which is available at https://www.rstudio.com/resources/cheatsheets/ and provides an overview of the most important **ggplot2** functions.

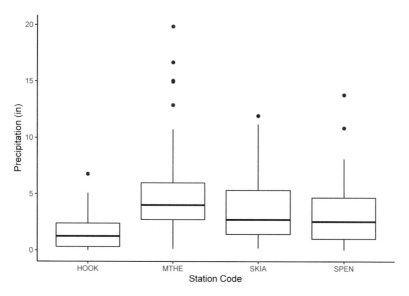

FIGURE 2.18
Boxplots of rainfall at four weather stations.

2.9 Practice

Write code to generate the following graphs based on the mesosm dataset. You can start with the examples provided in the text and modify them as needed.

1. Create a graph that displays four scatterplots of TMIN versus TMAX—one for each site.

2. Create a boxplot that compares the distribution of TMAX for each site. Make the axis text and labels bold so that they are easier to see.

3. Create a graph that displays four histograms of RAIN, one for each site. Experiment with changing the number of bins in the histograms to see how this affects the visualization.

3

Processing Tabular Data

The preceding chapter provided a foundation for visualizing data using the **ggplot2** package. The sample dataset used in the examples was already in a format suitable for plotting with `ggplot()`. However, in most cases, data need to be reformatted before they can be visualized and analyzed. Common formatting tasks include selecting subsets of rows and columns from the data table, calculating new variables from the raw data values, computing summary statistics, and combining data from different sources.

This tutorial will demonstrate basic functions for manipulating data frames (and tibbles). These manipulations can be accomplished in a variety of ways, including using base R operators and functions like those covered in Chapter 1. However, the functions in the **dplyr** and **tidyr** packages provide a more consistent and intuitive approach. These packages (along with the **ggplot2**, **readr**, and **readxl** packages used in Chapter 2) are part of the **tidyverse**, a collection of data science R packages. These packages share an underlying design philosophy, grammar of data manipulation, and set of data structures. The rest of this book will make extensive use of these packages. They will provide the basis for carrying out sophisticated processing, analysis, and visualization of complex "real-world" datasets by writing concise and easily interpretable R code.

To run the examples in this chapter, it is necessary to use the `library()` function to load the previously-mentioned **tidyverse** packages. Remember that if these packages are not available on your computer, you will need to install them using the `install.packages()` function or the installation tools available in RStudio.

```
library(ggplot2)
library(dplyr)
library(tidyr)
library(readr)
library(readxl)
```

The tutorials in this chapter will use the same Oklahoma Mesonet dataset that was used in Chapter 2. The output is printed to the console with the `glimpse()` function. This function shows the first few rows of data from mesosm,

DOI: 10.1201/9781003326199-3

with one column from the data frame shown on each row of the printed output. This format makes it easier to see all the variables in the data frame without having the wrap the console output across multiple lines.

```
mesosm <- read_csv("mesodata_small.csv", show_col_types = FALSE)
glimpse(mesosm)
## Rows: 240
## Columns: 9
## $ MONTH <dbl> 1, 2, 3, 4, 5, 6, 7, 8, 9, 10, 11, 12, 1, 2,~
## $ YEAR  <dbl> 2014, 2014, 2014, 2014, 2014, 2014, 2014, 20~
## $ STID  <chr> "HOOK", "HOOK", "HOOK", "HOOK", "HOOK", "HOO~
## $ TMAX  <dbl> 49.48032, 47.18071, 60.70613, 72.36483, 84.3~
## $ TMIN  <dbl> 17.92903, 17.05357, 26.06806, 39.32103, 48.3~
## $ HMAX  <dbl> 83.04161, 88.28857, 78.96613, 81.82690, 75.3~
## $ HMIN  <dbl> 29.00226, 39.94393, 25.39871, 21.02207, 18.8~
## $ RAIN  <dbl> 0.17, 0.30, 0.31, 0.40, 1.25, 3.18, 2.58, 0.~
## $ DATE  <date> 2014-01-01, 2014-02-01, 2014-03-01, 2014-04~
```

3.1 Single Table Verbs

Each **dplyr** function accomplishes a particular type of data transformation. These functions are referred to as "verbs" that describe the action performed. Some **dplyr** verbs operate on a single data frame or tibble, while others operate by combining data from two or more. Below is a list of some important **dplyr** verbs for single tables:

- select() and rename() select variables (columns) based on their names.
- filter() selects observations (rows) based on their values.
- arrange() reorders observations.
- mutate() and transmute() add new variables that are functions of existing variables.

The first argument to each function is the data frame that will be modified. Additional comma-separated arguments control how the function will be implemented.

3.1.1 Select and rename

In the context of **dplyr**, selecting specifically refers to choosing a subset of the columns within a data frame based on user-specified criteria. The simplest way is to specify the names of the columns to be retained. The select() function returns only the specified columns. Other columns are removed from the data

frame. One way to specify the selected columns is a comma-separated list of
column names. In the following examples, the outputs are not assigned to a
new object, so by default, they are printed to the console.

```
select(mesosm, STID, YEAR, MONTH, TMAX, TMIN)
## # A tibble: 240 x 5
##     STID   YEAR MONTH  TMAX   TMIN
##     <chr> <dbl> <dbl> <dbl>  <dbl>
##  1 HOOK   2014      1  49.5   17.9
##  2 HOOK   2014      2  47.2   17.1
##  3 HOOK   2014      3  60.7   26.1
##  4 HOOK   2014      4  72.4   39.3
##  5 HOOK   2014      5  84.4   48.3
##  6 HOOK   2014      6  90.7   61.9
##  7 HOOK   2014      7  90.6   64.7
##  8 HOOK   2014      8  95.8   64.5
##  9 HOOK   2014      9  84.3   58.0
## 10 HOOK   2014     10  76.1   44.9
## # ... with 230 more rows
```

The : operator can be used to select a continuous series of columns. This
usage is analogous to the : operator in base R, where it is used to specify a
continuous series of integers.

```
select(mesosm, MONTH:TMAX)
## # A tibble: 240 x 4
##     MONTH  YEAR STID   TMAX
##     <dbl> <dbl> <chr> <dbl>
##  1     1  2014 HOOK   49.5
##  2     2  2014 HOOK   47.2
##  3     3  2014 HOOK   60.7
##  4     4  2014 HOOK   72.4
##  5     5  2014 HOOK   84.4
##  6     6  2014 HOOK   90.7
##  7     7  2014 HOOK   90.6
##  8     8  2014 HOOK   95.8
##  9     9  2014 HOOK   84.3
## 10    10  2014 HOOK   76.1
## # ... with 230 more rows
```

The helper functions starts_with(), ends_with(), and contains() can be used
to find multiple columns by matching part of the column name.

```
select(mesosm, starts_with("T"))
## # A tibble: 240 x 2
##       TMAX   TMIN
##      <dbl>  <dbl>
##  1    49.5   17.9
##  2    47.2   17.1
##  3    60.7   26.1
##  4    72.4   39.3
##  5    84.4   48.3
##  6    90.7   61.9
##  7    90.6   64.7
##  8    95.8   64.5
##  9    84.3   58.0
## 10    76.1   44.9
## # ... with 230 more rows
```

Columns can be removed by prefixing their names with a -. Other columns will be kept.

```
select(mesosm, -HMIN, -HMAX)
## # A tibble: 240 x 7
##     MONTH  YEAR STID   TMAX  TMIN  RAIN DATE
##     <dbl> <dbl> <chr> <dbl> <dbl> <dbl> <date>
##  1      1  2014 HOOK   49.5  17.9  0.17 2014-01-01
##  2      2  2014 HOOK   47.2  17.1  0.3  2014-02-01
##  3      3  2014 HOOK   60.7  26.1  0.31 2014-03-01
##  4      4  2014 HOOK   72.4  39.3  0.4  2014-04-01
##  5      5  2014 HOOK   84.4  48.3  1.25 2014-05-01
##  6      6  2014 HOOK   90.7  61.9  3.18 2014-06-01
##  7      7  2014 HOOK   90.6  64.7  2.58 2014-07-01
##  8      8  2014 HOOK   95.8  64.5  0.95 2014-08-01
##  9      9  2014 HOOK   84.3  58.0  1.48 2014-09-01
## 10     10  2014 HOOK   76.1  44.9  1.72 2014-10-01
## # ... with 230 more rows
```

The rename() function is used to change column names. Name changes are specified using the = operator, placing the new name first and the old name second. After an external dataset has been imported, it is often desirable to change the names of the columns to make them shorter, more interpretable, or more consistent with other column names.

```
rename(mesosm, maxtemp = TMAX, mintemp = TMIN)
## # A tibble: 240 x 9
```

```
##     MONTH  YEAR STID  maxtemp mintemp  HMAX  HMIN  RAIN
##     <dbl> <dbl> <chr>   <dbl>   <dbl> <dbl> <dbl> <dbl>
##  1      1  2014 HOOK     49.5    17.9  83.0  29.0  0.17
##  2      2  2014 HOOK     47.2    17.1  88.3  39.9  0.3
##  3      3  2014 HOOK     60.7    26.1  79.0  25.4  0.31
##  4      4  2014 HOOK     72.4    39.3  81.8  21.0  0.4
##  5      5  2014 HOOK     84.4    48.3  75.4  18.8  1.25
##  6      6  2014 HOOK     90.7    61.9  90.9  28.8  3.18
##  7      7  2014 HOOK     90.6    64.7  88.2  33.1  2.58
##  8      8  2014 HOOK     95.8    64.5  85.4  21.8  0.95
##  9      9  2014 HOOK     84.3    58.0  91.2  36.4  1.48
## 10     10  2014 HOOK     76.1    44.9  85.7  28.1  1.72
## # ... with 230 more rows, and 1 more variable: DATE <date>
```

3.1.2   The pipe operator

Before covering additional functions, it is time to introduce a new operator called the *pipe*, %>%. Although it is possible to use **dplyr** and **tidyr** without the pipe operator, piping is a feature that helps to make coding with these packages more efficient and effective. When a pipe is placed on the right side of an object or function, the output from the function is passed as the first argument to the next function after the pipe. This is a simple example of using the pipe operator with the select function.

```
mesosm %>%
  select(MONTH:TMAX)
## # A tibble: 240 x 4
##     MONTH  YEAR STID   TMAX
##     <dbl> <dbl> <chr> <dbl>
##  1      1  2014 HOOK   49.5
##  2      2  2014 HOOK   47.2
##  3      3  2014 HOOK   60.7
##  4      4  2014 HOOK   72.4
##  5      5  2014 HOOK   84.4
##  6      6  2014 HOOK   90.7
##  7      7  2014 HOOK   90.6
##  8      8  2014 HOOK   95.8
##  9      9  2014 HOOK   84.3
## 10     10  2014 HOOK   76.1
## # ... with 230 more rows
```

The output from this code is the same as the output from select(mesosum, MONTH:TMAX), which was run in a previous example. The pipe indicates that mesosum should be used as the first argument to the select() function.

3.1.3   Filter

Filtering involves choosing a subset of rows from a data frame using criteria specified in one or more logical statements. These examples use the filter() function to select records by station, year, temperature, and humidity values. They are combined with the select() function using pipes so that only a subset of columns are printed to the console. This example selects only rows from the Mount Herman station, which has an ID code of "MTHE". The mesosm data frame is piped to become the first argument to the filter() function, and the data frame generated by filter() is piped to become the first argument to the select() function.

```
mesosm %>%
  filter(STID == "MTHE") %>%
  select(STID, MONTH, TMAX, HMAX)
## # A tibble: 60 x 4
##    STID  MONTH  TMAX  HMAX
##    <chr> <dbl> <dbl> <dbl>
##  1 MTHE      1  50.5  83.8
##  2 MTHE      2  50.9  89.2
##  3 MTHE      3  60.7  92.4
##  4 MTHE      4  69.9  91.2
##  5 MTHE      5  77.5  93.4
##  6 MTHE      6  84.4  94.7
##  7 MTHE      7  85.5  96.2
##  8 MTHE      8  88.6  95.6
##  9 MTHE      9  83.8  95.6
## 10 MTHE     10  75.6  96.3
## # ... with 50 more rows
```

The %in% operator returns TRUE if the input matches one or more of the values in the subsequent vector. The : operator is used here as a shortcut to create a vector containing only the months of June (month 6) through September (month 9), as demonstrated in Chapter 1.

```
mesosm %>%
  filter(MONTH %in% 6:9) %>%
  select(STID, MONTH, TMAX, HMAX)
## # A tibble: 80 x 4
##    STID  MONTH  TMAX  HMAX
```

```
##      <chr> <dbl> <dbl> <dbl>
##  1 HOOK      6  90.7  90.9
##  2 HOOK      7  90.6  88.2
##  3 HOOK      8  95.8  85.4
##  4 HOOK      9  84.3  91.2
##  5 HOOK      6  90.8  89.8
##  6 HOOK      7  94.6  88.6
##  7 HOOK      8  93.0  90.6
##  8 HOOK      9  90.7  84.8
##  9 HOOK      6  93.9  88.0
## 10 HOOK      7  98.5  86.7
## # ... with 70 more rows
```

When multiple logical statements are separated by commas, they are combined using a logical "and" operator. This example selects rows with high maximum temperature and high maximum relative humidity.

```
mesosm %>%
  filter(TMAX > 92, HMAX > 90) %>%
  select(STID, MONTH, TMAX, HMAX)
## # A tibble: 6 x 4
##    STID  MONTH  TMAX  HMAX
##   <chr> <dbl> <dbl> <dbl>
## 1 HOOK      8  93.0  90.6
## 2 HOOK      7  93.9  91.3
## 3 HOOK      8  92.1  90.6
## 4 MTHE      7  92.7  94.8
## 5 MTHE      7  93.8  92.4
## 6 SKIA      7  92.4  92.5
```

3.1.4   Arrange

Arranging changes the order of rows in a data frame based on the ranks of values in one or more columns. The `arrange()` function returns a data frame that is sorted on the comma-separated column names in order from left to right.

```
mesosm %>%
  arrange(MONTH, YEAR, STID) %>%
  select(MONTH, YEAR, STID)
## # A tibble: 240 x 3
##    MONTH  YEAR STID
##    <dbl> <dbl> <chr>
```

```
##  1       1  2014  HOOK
##  2       1  2014  MTHE
##  3       1  2014  SKIA
##  4       1  2014  SPEN
##  5       1  2015  HOOK
##  6       1  2015  MTHE
##  7       1  2015  SKIA
##  8       1  2015  SPEN
##  9       1  2016  HOOK
## 10       1  2016  MTHE
## # ... with 230 more rows
```

The desc() function can be used to order a column in descending rather than ascending order.

```
mesosm %>%
  arrange(MONTH, desc(YEAR), desc(STID)) %>%
  select(MONTH, YEAR, STID)
## # A tibble: 240 x 3
##     MONTH  YEAR STID
##     <dbl> <dbl> <chr>
##  1      1  2018  SPEN
##  2      1  2018  SKIA
##  3      1  2018  MTHE
##  4      1  2018  HOOK
##  5      1  2017  SPEN
##  6      1  2017  SKIA
##  7      1  2017  MTHE
##  8      1  2017  HOOK
##  9      1  2016  SPEN
## 10      1  2016  SKIA
## # ... with 230 more rows
```

3.1.5   Mutate and transmute

Mutate and transmute are used to add new columns derived from the values in existing columns. The mutate() function retains all the columns in the input data frame and adds new columns. Multiple new variables can be generated with a single function call using a comma to separate each new variable. The name of each new variable is specified on the left of the = operator, and a function that can contain the names of other columns in the data frame is specified on the right. The following example converts the minimum and maximum temperature variables from Fahrenheit to Celsius.

```
mesosm %>%
  mutate(TMINC = (TMIN - 32) * .5556,
         TMAXC = (TMAX - 32) * .5556) %>%
  select(MONTH, YEAR, STID, TMIN, TMAX, TMINC, TMAXC)
## # A tibble: 240 x 7
##    MONTH  YEAR STID   TMIN  TMAX TMINC TMAXC
##    <dbl> <dbl> <chr> <dbl> <dbl> <dbl> <dbl>
##  1     1  2014 HOOK  17.9  49.5 -7.82  9.71
##  2     2  2014 HOOK  17.1  47.2 -8.30  8.43
##  3     3  2014 HOOK  26.1  60.7 -3.30 15.9
##  4     4  2014 HOOK  39.3  72.4  4.07 22.4
##  5     5  2014 HOOK  48.3  84.4  9.06 29.1
##  6     6  2014 HOOK  61.9  90.7 16.6  32.6
##  7     7  2014 HOOK  64.7  90.6 18.2  32.6
##  8     8  2014 HOOK  64.5  95.8 18.0  35.4
##  9     9  2014 HOOK  58.0  84.3 14.4  29.0
## 10    10  2014 HOOK  44.9  76.1  7.17 24.5
## # ... with 230 more rows
```

In contrast, the `transmute()` function only includes the newly-created variables in the output.

```
mesosm %>% transmute(TMINC = (TMIN - 32) * .5556,
                     TMAXC = (TMIN - 32) * .5556)
## # A tibble: 240 x 2
##    TMINC TMAXC
##    <dbl> <dbl>
##  1 -7.82 -7.82
##  2 -8.30 -8.30
##  3 -3.30 -3.30
##  4  4.07  4.07
##  5  9.06  9.06
##  6 16.6  16.6
##  7 18.2  18.2
##  8 18.0  18.0
##  9 14.4  14.4
## 10  7.17  7.17
## # ... with 230 more rows
```

As shown in these examples, the `mutate()` and `transmute()` functions can be used to create multiple new columns. In the example code, hard returns after each comma have been used to distribute each function call over multiple lines of code. A similar approach was used in Chapter 3 when writing code for `ggplot()` functions with multiple arguments. This is a common programming technique that can make the code more interpretable. Many programmers find that a multi-line format with one argument per line is easier to read for long and complex function calls. Formatting a function across multiple lines does not affect how the function is executed in R. This book uses both single-line and multi-line formats for code depending on the context. In most cases, functions with one or two arguments are coded on single lines, and functions with three or more arguments are broken across multiple lines.

3.1.6   Application

The following examples will use another meteorological dataset from the Oklahoma Mesonet. The `mesodata_large.csv` file contains daily data records from every Mesonet station in Oklahoma from 1994 to 2018. This is a very large data table with more than one million rows and over 150 Mb of data. There are also numerous missing data codes (values $<= -990$ or $>= 990$) that need to be dealt with.

```
mesobig <- read_csv("mesodata_large.csv", show_col_types = FALSE)
dim(mesobig)
## [1] 1296602       22
names(mesobig)
##  [1] "YEAR"   "MONTH" "DAY"   "STID"   "TMAX"   "TMIN"   "TAVG"
##  [8] "DMAX"   "DMIN"  "DAVG"  "HMAX"   "HMIN"   "HAVG"   "VDEF"
## [15] "9AVG"   "HDEG"  "CDEG"  "HTMX"   "WCMN"   "RAIN"   "RNUM"
## [22] "RMAX"
```

The goal is to create a smaller dataset that is identical to the `mesosm` dataset that was introduced earlier. This dataset should contain monthly summaries of temperature, precipitation, and humidity for the Hooker, Mt. Herman, Skiatook, and Spencer stations.

The **dplyr** single-table functions can be combined to create this smaller version of the big Mesonet dataset. The first step is to filter out records (rows) that belong to one of four sites and are from the years 2014 to present. The output is saved to a data frame object called `mesonew`. Then, the `select()` function is used to choose a subset of the columns. Finally, the temperature, humidity, and rainfall records are transformed by replacing the missing data codes (values less than or equal to -990 and greater than or equal to 990) with `NA` values using the `mutate()` and `replace()` functions. The `replace()` function takes three comma-separated arguments. The first is the variable to replace, the second is a logical statement indicating which values will be replaced, and the third is the replacement value. In this example, the columns are overwritten with their updated values rather than creating new columns.

```
mesonew <- mesobig %>%
  filter(STID %in% c("HOOK", "SKIA", "MTHE", "BUTL")
         & YEAR >= 2014) %>%
  select(STID, YEAR, MONTH, DAY, TMIN, TMAX, HMIN, HMAX, RAIN) %>%
  mutate(TMIN = replace(TMIN, TMIN <= -990 | TMIN >= 990, NA),
         TMAX = replace(TMAX, TMAX <= -990 | TMAX >= 990, NA),
         HMIN = replace(HMIN, HMIN <= -990 | HMIN >= 990, NA),
         HMAX = replace(HMAX, HMAX <= -990 | HMAX >= 990, NA),
         RAIN = replace(RAIN, RAIN <= -990 | RAIN >= 990, NA))
```

The assignment operator on the first line of this code block saves the output of
these combined functions as a new object called mesonew. One of the advantages
of using pipes is that it is not necessary to explicitly save the output of each
function as a new object and then specify that object as an argument to the
next function. The resulting code is thus more concise and easier to read. In
most cases, there is no need to save the intermediate outputs of every step
because all of the code can easily be rerun if a change is required or an error
needs to be fixed.

3.2   Summarizing

At this point, the numbers of rows and columns in the data frame has been
reduced. The missing data codes have been removed and replaced with NA
values.

```
dim(mesonew)
## [1] 7304    9
names(mesonew)
## [1] "STID"   "YEAR"   "MONTH" "DAY"    "TMIN"   "TMAX"   "HMIN"
## [8] "HMAX"   "RAIN"
summary(mesonew)
##      STID                 YEAR          MONTH
##  Length:7304       Min.   :2014   Min.   : 1.000
##  Class :character  1st Qu.:2015   1st Qu.: 4.000
##  Mode  :character  Median :2016   Median : 7.000
##                    Mean   :2016   Mean   : 6.524
##                    3rd Qu.:2017   3rd Qu.:10.000
##                    Max.   :2018   Max.   :12.000
##
##       DAY              TMIN              TMAX
```

```
## Min.    : 1.00   Min.    :-16.15   Min.    :  6.78
## 1st Qu.: 8.00   1st Qu.: 32.99   1st Qu.: 59.69
## Median :16.00   Median : 49.30   Median : 74.66
## Mean   :15.73   Mean    : 47.78   Mean    : 72.25
## 3rd Qu.:23.00   3rd Qu.: 64.36   3rd Qu.: 87.19
## Max.   :31.00   Max.    : 79.32   Max.    :110.59
##                 NA's    :9        NA's    :9
##         HMIN            HMAX             RAIN
## Min.    :  2.46   Min.    : 27.13   Min.    :0.0000
## 1st Qu.: 26.86   1st Qu.: 84.80   1st Qu.:0.0000
## Median : 38.26   Median : 93.01   Median :0.0000
## Mean    : 41.23   Mean    : 89.47   Mean    :0.1053
## 3rd Qu.: 53.23   3rd Qu.: 97.80   3rd Qu.:0.0100
## Max.   :100.00   Max.    :100.00   Max.    :8.8300
## NA's    :62       NA's    :62       NA's    :9
```

However, there is still one record in the data frame for each day. The next step is to convert these daily values to monthly summaries. Summarizing a data frame involves calculating summary statistics for one or more columns across multiple rows. This is accomplished using the summarize() function. The following example calculates the mean maximum temperature for all records in the data frame. As discussed in Chapter 1, the na.rm = TRUE argument is specified so that the mean() function will ignore the NA values.

```
mesonew %>%
  summarize(meantmax = mean(TMAX, na.rm = TRUE))
## # A tibble: 1 x 1
##   meantmax
##      <dbl>
## 1     72.2
```

The summarize() function is particularly useful when paired with another dplyr function, group_by(), which groups rows of data together to produce a grouped data frame. After a data frame has been grouped, it belongs to the class grouped_df.

```
mesogrp <- group_by(mesonew, STID, YEAR, MONTH)
class(mesogrp)
## [1] "grouped_df" "tbl_df"    "tbl"       "data.frame"
```

When a grouped data frame is summarized, the summaries are generated for individual groups rather than the entire dataset, and in this case the result is a data frame with one row for each combination of station, year, and month. Grouping is an intermediate step toward summarizing by group, and there is usually no need to save the grouped data frame. Therefore, grouping and summarization are often combined using pipes. In this example, the input is the `mesonew` dataset that was already created, and the output is a new data frame called `mesomnth`

```
mesomnth <- mesonew %>%
  group_by(STID, YEAR, MONTH) %>%
  summarise(TMAX = mean(TMAX, na.rm=T),
            TMIN = mean(TMIN, na.rm=T),
            HMAX = mean(HMAX, na.rm=T),
            HMIN = mean(HMIN, na.rm=T),
            RAIN = sum(RAIN, na.rm=T))
## `summarise()` has grouped output by 'STID', 'YEAR'. You can
## override using the `.groups` argument.
```

Finally, the `mutate()` function is used to add a date column. Because each record represents an entire month, the date is set to correspond to the first day of each month. The `as.Date()` function is used to create the date object. The input to this function is a date string with year followed by month followed by day separated by dashes. For example, March 1, 2018 would be "2018-3-1." The `paste()` function joins multiple strings using a separator specified with the `sep` argument. Because these are monthly data, there is only a `YEAR` and `MONTH` value for each record. The date associated with each month is assumed to be the first day of the month, and a value of "1" is therefore used for the day. More details on working with dates will be provided in Chapter 4.

```
mesomnth <- mutate(mesomnth,
                   DATE = as.Date(paste(YEAR,
                                        MONTH,
                                        "1",
                                        sep = "-")))
```

The `glimpse()` function provides an overview of all the columns in the new `mesomnth` data frame with examples of the data values.

```
glimpse(mesomnth)
## Rows: 240
## Columns: 9
## Groups: STID, YEAR [20]
## $ STID  <chr> "BUTL", "BUTL", "BUTL", "BUTL", "BUTL", "BUT~
## $ YEAR  <dbl> 2014, 2014, 2014, 2014, 2014, 2014, 2014, 20~
## $ MONTH <dbl> 1, 2, 3, 4, 5, 6, 7, 8, 9, 10, 11, 12, 1, 2,~
## $ TMAX  <dbl> 52.02516, 47.52321, 61.08935, 74.87600, 84.7~
## $ TMIN  <dbl> 21.58161, 22.25393, 31.26387, 45.34533, 56.0~
```

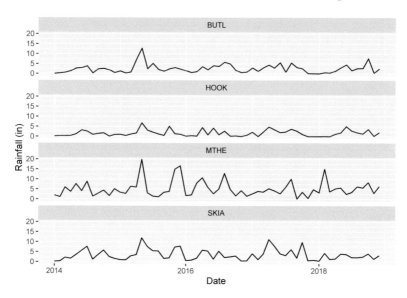

FIGURE 3.1
Faceted line graph of monthly rainfall generated with the mesomnth dataset.

```
## $ HMAX   <dbl> 76.27871, 88.79679, 79.48548, 80.23667, 79.1~
## $ HMIN   <dbl> 27.18097, 41.16464, 27.06161, 25.08967, 28.3~
## $ RAIN   <dbl> 0.01, 0.26, 0.59, 1.33, 2.76, 2.97, 3.85, 0.~
## $ DATE   <date> 2014-01-01, 2014-02-01, 2014-03-01, 2014-04~
```

You can also look at the resulting mesomnth data frame in RStudio by running
the View(mesomnth) function. The data frame will open up as a new tab in the
same window that contains your R scripts. You can easily scroll through the
rows and across the columns to view the raw data values. They should
be the same as the mesosm dataset that was used in Chapter 2 and imported at
the beginning of this chapter, and they can be used to generate similar graphs
(Figure 3.1).

```
ggplot(data = mesomnth) +
  geom_line(mapping = aes(x = DATE, y = RAIN)) +
  facet_wrap(~ STID, ncol = 1) +
  labs(x = "Date",
      y = "Rainfall (in)",
      color = "Station ID")
```

3.2.1 Counts

When summarizing data, it is often necessary to count the number of values used. This can be accomplished with the **dplyr** function n(), which will include NA values in the count, or sum(!is.na(x)) which will exclude them. It's a good idea to count data when summarizing them to see the sample sizes that are being used to calculate the summary statistics for each group.

This example uses the mesonew data from the previous section. The data frame is first grouped by location, year, and month. Several summary variables are then calculated with summarize(). The new n_rows column is a count of all observations for each group. The obs_hmax column is a count of the number of rows that do *not* have missing data for maximum humidity. The pv_hmax column is calculated as the percentage of observations that have valid maximum humidity data for each group. The mutate() function is used to add a date column to the mesomissing data frame.

```
mesomissing <- mesonew %>%
  group_by(STID, YEAR, MONTH) %>%
  summarize(n_rows = n(),
            obs_hmax = sum(!is.na(HMAX)),
            pv_hmax = 100 * obs_hmax/n_rows) %>%
  mutate(DATE = as.Date(paste(YEAR, MONTH, "1", sep = "-")))
glimpse(mesomissing)
## Rows: 240
## Columns: 7
## Groups: STID, YEAR [20]
## $ STID     <chr> "BUTL", "BUTL", "BUTL", "BUTL", "BUTL", "~
## $ YEAR     <dbl> 2014, 2014, 2014, 2014, 2014, 2014, 2014,~
## $ MONTH    <dbl> 1, 2, 3, 4, 5, 6, 7, 8, 9, 10, 11, 12, 1,~
## $ n_rows   <int> 31, 28, 31, 30, 31, 30, 31, 31, 30, 31, 3~
## $ obs_hmax <int> 31, 28, 31, 30, 31, 30, 31, 31, 30, 31, 3~
## $ pv_hmax  <dbl> 100, 100, 100, 100, 100, 100, 100, 100, 1~
## $ DATE     <date> 2014-01-01, 2014-02-01, 2014-03-01, 2014~
```

The resulting data frame contains information about when and where there are missing data. An efficient way to explore these data is using a simple plot. The following example generates a line plot, similar to the examples in Chapter 2, that displays the monthly time series of missing data using a separate plot for each site (Figure 3.2).

```
ggplot(data = mesomissing) +
  geom_line(mapping = aes(x = DATE, y = pv_hmax)) +
  facet_wrap(~ STID, ncol = 1) +
  labs(x = "Date",
```

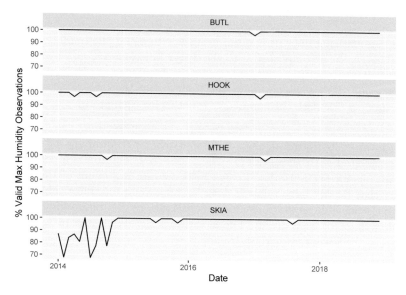

FIGURE 3.2
Line graphs of valid humidity observations for four stations.

```
y = "% Valid Max Humidity Observations",
color = "Station ID")
```

3.2.2   Summary functions

Up to this point, only a few summary functions have been covered: `sum()`, `mean()`, and `n()`. Many other summary functions are available including:

- Measures of central tendency, including the `median()` as well as the `mean()`.
- Measure of variability, including `sd()`, the standard deviation; `IQR()`, the inter-quartile range; and `mad()`, the mean absolute deviation.
- Measures of rank, including `min()`, the minimum value; `quantile()`, the data value associated with a given percentile; and `max()`, the maximum value.
- Counts, including `n()`, which counts all rows; and `n_distinct()`, which counts the number of unique values.

3.3   Pivoting Data

Almost all of the data used in environmental geography applications can be stored and manipulated in data frames, but the format of the data frame can differ. Consider the Mesonet data where each observation is indexed by a location (station) and a time (month and year). Each weather variable, such as rainfall, is stored in a separate column. The rainfall at a particular location and time can be determined by referencing the appropriate row based on the values in the station, month, and year columns. This is what is called a "long" format, which minimizes the number of columns by increasing the number of rows in the dataset.

These data could also be stored in a "wide" format. For example, rainfall data from each station could be stored as a separate column with one row for each combination of month and year. The rainfall at a particular location and time would then be determined by referencing the appropriate column for a given station and the appropriate row based on values in the month and year columns.

In general, the "long" format is considered to be a more efficient and "tidy" way to format and process data. This idea of "tidy" data is the conceptual foundation for the **tidyr** package as well as the larger **tidyverse** collection of packages. In most cases, a long data format with fewer columns and more rows is easiest to manipulate using **dplyr** functions and is the format needed for plotting with `ggplot()` and running many types of statistical models. However, there are other situations where a wider format can be more suitable for data manipulation and analysis. Therefore the **tidyr** package provides two important functions called `pivot_longer()` and `pivot wider()` that can be used to reformat data frames.

Consider the problem of plotting both minimum and maximum temperatures on the same graph with `ggplot()`. In the current `mesomnth` data frame, they are in separate columns - `TMIN` and `TMAX`. To plot them in the same graph with different aesthetics, it is helpful to have all of the temperature observations in a single *values* column and to have a second *names* column that indicates the type of temperature measurement (minimum or maximum). The `pivot_longer()` function takes three basic arguments. The `cols` argument specifies which columns from the input dataset will be combined into the new values column. The `one_of()` function is used to specify multiple column names from the input data frame. The `values_to` argument specifies the name of the new column that will hold the data values. The `names_to` argument specifies the column with the new variable names. The `mutate()` function is also used to convert `tstat` from a character variable to a labeled factor.

```
temp_tidy <- mesomnth %>%
  pivot_longer(cols = one_of("TMAX", "TMIN"),
               values_to = "temp",
               names_to = "tstat") %>%
  mutate(tstat = factor(tstat,
                        levels = c("TMAX", "TMIN"),
```

```
                             labels = c("Maximum", "Minimum")))
glimpse(temp_tidy)
## Rows: 480
## Columns: 9
## Groups: STID, YEAR [20]
## $ STID  <chr> "BUTL", "BUTL", "BUTL", "BUTL", "BUTL", "BUT~
## $ YEAR  <dbl> 2014, 2014, 2014, 2014, 2014, 2014, 2014, 20~
## $ MONTH <dbl> 1, 1, 2, 2, 3, 3, 4, 4, 5, 5, 6, 6, 7, 7, 8,~
## $ HMAX  <dbl> 76.27871, 76.27871, 88.79679, 88.79679, 79.4~
## $ HMIN  <dbl> 27.18097, 27.18097, 41.16464, 41.16464, 27.0~
## $ RAIN  <dbl> 0.01, 0.01, 0.26, 0.26, 0.59, 0.59, 1.33, 1.~
## $ DATE  <date> 2014-01-01, 2014-01-01, 2014-02-01, 2014-02~
## $ tstat <fct> Maximum, Minimum, Maximum, Minimum, Maximum,~
## $ temp  <dbl> 52.02516, 21.58161, 47.52321, 22.25393, 61.0~
```

Now there is only a single column with temperature measurements (temp), and another column with a code that indicates the type of temperature measurement (tstat). However, the temp_tidy data frame has twice as many rows as the original mesomnth data frame because each minimum and maximum temperature value occupies a separate row. The tstat column is now a factor, and printing the data frame with glimpse() shows the more interpretable class labels (Maximum and Minimum) instead of the original variable codes (TMAX and TMIN).

Organizing the data in this long format makes it easier to generate a graph with ggplot() that includes minimum and maximum temperatures (Figure 3.3). The labels of the tstat factor are shown in the legend, and the legend title is modified using the labs() function.

```
ggplot(temp_tidy) +
  geom_line(aes(x = DATE,
                y = temp,
                color = tstat)) +
  facet_wrap(facets = vars(STID),
             ncol = 1,
             scales = "free_y") +
  labs(color = "Temperature",
       x = "Date",
       y = "Temperature (\u00B0F)")
```

When observations are distributed across multiple rows, it is sometimes necessary to reorganize the data into a wide format with multiple columns. For example, consider the problem of generating a table with just one row for each location and columns containing the total rainfall for each month in

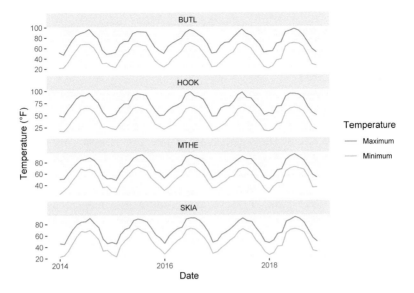

FIGURE 3.3
Line graph of monthly minimum and maximum temperatures at four stations.

2018. The code below shows how to accomplish this task using several **dplyr** functions. The data are filtered to include only records from 2018, the necessary columns are selected, missing data codes are replaced by NA, and total rainfall is calculated for each month at each weather station. The **tidyr** `pivot_wider()` function is then used to create a separate column for each monthly rainfall summary. The `values_from` argument indicates the input column with data to be distributed over multiple output columns. The names of these new columns will be generated by pasting the `names_prefix` string to the values in the `names_from` column.

```
meso2018 <- mesobig %>%
  filter(YEAR == 2018) %>%
  select(STID, MONTH, RAIN) %>%
  mutate(RAIN = replace(RAIN, RAIN <= -990 | RAIN >= 990, NA)) %>%
  group_by(STID, MONTH) %>%
  summarise(RAIN = sum(RAIN, na.rm=T)) %>%
  pivot_wider(values_from = RAIN,
              names_from = MONTH,
              names_prefix = "M")
```

The `meso2018` data frame now has only 2018 data with one row for each station. Twelve columns named M1-M12 contain the monthly rainfall observations from January through December.

```
glimpse(meso2018)
## Rows: 142
## Columns: 13
## Groups: STID [142]
## $ STID  <chr> "ACME", "ADAX", "ALTU", "ALV2", "ALVA", "ANT2~
## $ M1    <dbl> 0.13, 0.37, 0.01, 0.00, 0.00, 1.33, 0.00, 0.1~
## $ M2    <dbl> 2.12, 7.50, 0.79, 0.34, 0.00, 10.31, 0.00, 2.~
## $ M3    <dbl> 0.67, 3.91, 0.54, 1.58, 0.00, 2.67, 0.00, 0.7~
## $ M4    <dbl> 1.80, 2.97, 0.97, 1.55, 0.00, 2.93, 0.00, 1.2~
## $ M5    <dbl> 4.77, 4.20, 3.62, 9.17, 0.00, 3.56, 0.00, 4.3~
## $ M6    <dbl> 4.25, 4.81, 2.24, 4.32, 0.00, 4.70, 0.00, 4.3~
## $ M7    <dbl> 3.04, 2.36, 1.31, 3.25, 0.00, 2.99, 0.00, 2.0~
## $ M8    <dbl> 1.71, 7.11, 7.05, 4.08, 0.00, 5.93, 0.00, 1.1~
## $ M9    <dbl> 10.19, 11.37, 3.51, 4.72, 0.00, 6.30, 0.00, 7~
## $ M10   <dbl> 6.61, 7.42, 8.60, 8.18, 0.00, 7.48, 0.00, 5.5~
## $ M11   <dbl> 0.54, 0.40, 0.24, 0.41, 0.00, 1.90, 0.00, 0.4~
## $ M12   <dbl> 5.44, 5.92, 1.09, 2.58, 0.00, 7.65, 0.00, 3.5~
```

3.4   Joining Tables

Table joins are a concept shared across many data science disciplines and
are implemented in relational database management systems such as MySQL.
With a join, two tables are connected to each other through variables called
keys, which are columns found in both tables. For the Mesonet data, possible
key columns include STID, YEAR, or MONTH.

To map the mesonet summaries, we will need to add information about the
geographic coordinates of each station to the summary table. This information
is in a separate dataset that contains, among other variables, the station ID
codes and the latitude and longitude of each weather station. There are several
join functions available in dplyr. All of these functions take tables as arguments,
which we will call x and y for simplicity. However, each function operates a bit
differently.

- inner_join(): returns all rows from x that have matching key variables in y
  and all columns from x and y. If there are multiple matches between x and y,
  all combinations of the matches are returned.
- left_join(): returns all rows from x and all columns from x and y. Rows
  in x with matching key variables in y will contain data in the y columns,
  and rows with no match in y will have NA values in the y columns. If there
  are multiple matches between x and y, all combinations of the matches are
  returned.

- `right_join()`: returns all rows from y and all columns from x and y. Rows in y with matching key variables in x will contain data in the x columns, and rows with no match in x will have NA values in the x columns. If there are multiple matches between x and y, all combinations of the matches are returned.
- `full_join()`: returns all rows and all columns from both x and y. Where there are no matching values, the function returns NA for columns in the unmatched table.

In this example, the `inner_join()` function is used to join the geographic coordinates to the existing meso2018 table. Both datasets have a station ID column, but the column name is in upper case in `meso2018` and in lower case in `geo_coords`. Specifying the by argument as `c("STID" = "stid")` indicates that these are the key columns that need to be matched from the x and y tables. In the call to `inner_join()`, the x table is the `meso2018` data frame that comes through the pipe, and the y table is the `geo_coords` data frame specified as the first function argument.

```
geo_coords <- read_csv("geoinfo.csv", show_col_types = FALSE)
mesospatial <- meso2018 %>%
  inner_join(geo_coords, by = c("STID" = "stid"))
mesospatial
## # A tibble: 142 x 24
## # Groups:   STID [142]
##     STID    M1    M2    M3    M4    M5    M6    M7    M8
##     <chr> <dbl> <dbl> <dbl> <dbl> <dbl> <dbl> <dbl> <dbl>
##  1 ACME   0.13  2.12  0.67  1.8   4.77  4.25  3.04  1.71
##  2 ADAX   0.37  7.5   3.91  2.97  4.2   4.81  2.36  7.11
##  3 ALTU   0.01  0.79  0.54  0.97  3.62  2.24  1.31  7.05
##  4 ALV2   0     0.34  1.58  1.55  9.17  4.32  3.25  4.08
##  5 ALVA   0     0     0     0     0     0     0     0
##  6 ANT2   1.33 10.3   2.67  2.93  3.56  4.7   2.99  5.93
##  7 ANTL   0     0     0     0     0     0     0     0
##  8 APAC   0.19  2.21  0.71  1.28  4.32  4.38  2     1.14
##  9 ARD2   0.12  7.31  2.36  1.93  6.84  1.45  0.9   4.26
## 10 ARDM   0     0     0     0     0     0     0     0
## # ... with 132 more rows, and 15 more variables: M9 <dbl>,
## #   M10 <dbl>, M11 <dbl>, M12 <dbl>, stnm <dbl>,
## #   name <chr>, city <chr>, rang <dbl>, cdir <chr>,
## #   cnty <chr>, lat <dbl>, lon <dbl>, elev <dbl>,
## #   cdiv <chr>, clas <chr>
```

The `mesospatial` data frame now contains monthly rainfall summaries and geographic coordinates for each Mesonet station, so `ggplot()` can be used to generate a simple map of January rainfall by plotting points with the longitudes

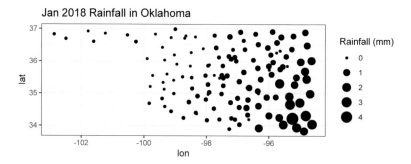

FIGURE 3.4
Graduated symbol map of January 2018 rainfall in Oklahoma.

on the x-axis, latitudes on the y-axis, and sizes scaled by January rainfall (Figure 3.4). Chapter 5 will build upon this simple example and introduce new object classes and functions for manipulating and mapping geospatial data.

```
ggplot(data = mesospatial) +
  geom_point(aes(x = lon,
                 y = lat,
                 size = M1),
             color = "black") +
  scale_size_continuous(name="Rainfall (mm)") +
  coord_equal() +
  labs(title="Jan 2018 Rainfall in Oklahoma") +
  theme_bw()
```

There are many helpful resources that can help expand your knowledge of **dplyr**, **tidyr**, and other **tidyverse** packages. Wickham and Grolemund's *R For Data Science* (Wickham and Grolemund, 2016) provides a comprehensive introduction to tabular data processing, visualization, and analysis. The website at https://www.tidyverse.org/ has helpful reference guides for all of the **tidyverse** packages. The "cheatsheets" available for **dplyr** and **tidyr** at https://www.rstudio.com/resources/cheatsheets/ are useful quick reference guides.

3.5   Practice

1. Write code to filter the rows from the Skiatook station (SKIA) where the precipitation is greater than one inch using the `mesosm` dataset.

2. Write code to select only the month, year, station ID, and rainfall columns using the mesosm dataset.

3. Write code to create a new column to the mesosm dataset called RAINMM that contains the rainfall values converted from inches to millimeters (1 mm = 0.3937 in).

4. Use the mesobig dataset to determine the percent of valid rainfall observations for each combination of station, month, and year and generate a graph of the results.

5. Use the mesosm dataset to generate a data frame containing minimum and maximum humidity in a long format and graph the results.

4

Dates in R

Several examples that use `Date` objects have already been shown in the preceding chapters. Dates are important in geographical data science because most environmental measurements are repeated over time. For example, automated weather stations can collect meteorological data on hourly or shorter measurement cycles, and Earth-observing satellites acquire images at a given location on daily to weekly cycles. These repeated measurements provide the basis for analyzing environmental change. When working with these data in R, it is essential to have information about the timing of observations as well as their locations in geographic space. This chapter will provide an overview of how to work with dates in R using the *lubridate* package (Spinu et al., 2021).

```
library(ggplot2)
library(dplyr)
library(readr)
library(lubridate)
```

4.1   Converting Characters to Dates

The `testdates.csv` file contains daily temperature measurements from an Oklahoma Mesonet station. The date of each observation is provided in a variety of formats, and these variables will be used to learn about manipulating dates with **lubridate**. Figure 4.1 shows the formats of the raw data in `testdates.csv`. The first row contains the column names, and the subsequent rows contain the data with each variable separated by a comma.

The columns contain the following information.

- YEAR: Year of meteorological observation
- STID: Meteorological station ID code
- MONTH: Month of meteorological observation
- DAY: Day of month of meteorological observation
- TAVG: Average temperature for date of observation
- DATE: Date formatted as "2018-01-09" (YYYY-MM-DD with leading zeroes)

DOI: 10.1201/9781003326199-4

```
YEAR,STID,MONTH,DAY,TAVG,DATE,DATE1,DATE2,DATE3
2018,COOK,1,1,11.5,2018-01-01,2018-1-1,1-1-2018,"January 1, 2018"
2018,COOK,1,2,16.86,2018-01-02,2018-1-2,1-2-2018,"January 2, 2018"T
2018,COOK,1,3,21.19,2018-01-03,2018-1-3,1-3-2018,"January 3, 2018"
2018,COOK,1,4,24.56,2018-01-04,2018-1-4,1-4-2018,"January 4, 2018"
2018,COOK,1,5,34.11,2018-01-05,2018-1-5,1-5-2018,"January 5, 2018"
2018,COOK,1,6,36.92,2018-01-06,2018-1-6,1-6-2018,"January 6, 2018"
2018,COOK,1,7,38.61,2018-01-07,2018-1-7,1-7-2018,"January 7, 2018"
2018,COOK,1,8,39.27,2018-01-08,2018-1-8,1-8-2018,"January 8, 2018"
2018,COOK,1,9,38.3,2018-01-09,2018-1-9,1-9-2018,"January 9, 2018"
2018,COOK,1,10,50.7,2018-01-10,2018-1-10,1-10-2018,"January 10, 2018"
2018,COOK,1,11,40.9,2018-01-11,2018-1-11,1-11-2018,"January 11, 2018"
2018,COOK,1,12,21.33,2018-01-12,2018-1-12,1-12-2018,"January 12, 2018"
```

FIGURE 4.1
Contents of the testdates.csv file.

- DATE1: Date formatted as "2018-1-9" (YYYY-MM-DD with no leading zeroes)
- DATE2: Date formatted as "1-9-2018" (MM-DD-YYYY with no leading zeroes)
- DATE3: Date formatted as "January 9, 2018" (MONTHNAME, DD, YYYY with no leading zeroes)

```
datefile <- read_csv("testdates.csv", show_col_type = FALSE)
glimpse(datefile)
## Rows: 365
## Columns: 9
## $ YEAR  <dbl> 2018, 2018, 2018, 2018, 2018, 2018, 2018, 20~
## $ STID  <chr> "COOK", "COOK", "COOK", "COOK", "COOK", "COO~
## $ MONTH <dbl> 1, 1, 1, 1, 1, 1, 1, 1, 1, 1, 1, 1, 1, 1, 1,~
## $ DAY   <dbl> 1, 2, 3, 4, 5, 6, 7, 8, 9, 10, 11, 12, 13, 1~
## $ TAVG  <dbl> 11.50, 16.86, 21.19, 24.56, 34.11, 36.92, 38~
## $ DATE  <date> 2018-01-01, 2018-01-02, 2018-01-03, 2018-01~
## $ DATE1 <chr> "2018-1-1", "2018-1-2", "2018-1-3", "2018-1-~
## $ DATE2 <chr> "1-1-2018", "1-2-2018", "1-3-2018", "1-4-201~
## $ DATE3 <chr> "January 1, 2018", "January 2, 2018", "Janua~
```

Note that read_csv() correctly interprets the DATE column and converts it to a Date object in R. The other date formats remain as character strings. The DATE column contains values formatted as YYYY-MM-DD. For example, January 1, 2018 is specified as 2018-01-01. There are always two digits for the month and day of the month, with leading zeroes included when these values are less than ten. The read_csv() function is able to recognize this format and automatically convert it to a date. The DATE1 and DATE2 columns are specified as YYYY-MM-DD and MM-DD-YYYY, and these columns do not include leading zeroes. Because these formats are more ambiguous, the dates cannot be automatically interpreted, and these columns are read as character strings. The same is true for the DATE3 column that spells out the month name.

Environmental datasets use a wide variety of date formats. Years, months, and days can be arranged in different orders using different separators. Months may be spelled out or numbered. Single-digit months and days may or may not have leading zeroes. In some cases, the month and day of the month may be replaced by a single number representing the day of the year.

These differences are especially problematic when data must be imported from multiple sources and joined by date. Furthermore, it is not uncommon to find inconsistently formatted date strings within a single dataset. Variables representing dates should always be carefully screened when importing a dataset, and extra attention is warranted to ensure that code is written to process the date variables correctly.

A `Date` object in R is stored as an integer value that represents the number of days before or after the baseline date of January 1, 1970. The following example generates a vector of date objects and shows their corresponding numeric values.

```
mydates <- as.Date(c("1932-01-01", "1950-12-25", "1968-04-28",
                "1990-06-24", "2003-07-04", "2033-11-15"))
as.numeric(mydates)
## [1] -13880  -6947   -613   7479  12237  23329
```

Other operators and functions in R are able to recognize the `Date` class, and they can extract and manipulate other types of information (years, months, weeks, days) associated with dates. For example, if we plot the DATE column on the x-axis of a graph, `ggplot()` knows how to handle it, and we get a time series plot of temperature (Figure 4.2). Tick marks on the x-axis are referenced by month and year.

```
ggplot(data = datefile) +
  geom_line(aes(x = DATE, y = TAVG)) +
  labs(x = "Correct Date",
       y = "Temperature (\u00B0F)")
```

The `DATE1`, `DATE2`, and `DATE3` columns all contain dates stored as character objects. If we tried to use one of these columns to make a time series graph in `ggplot()`, they would be interpreted as discrete categories instead of continuous time variables. Fortunately, it is relatively easy to convert character strings into dates. Base R has an `as.Date()` function that can convert some types of strings into `Date` objects.

```
datefile <- datefile %>%
  mutate(DATE1B = as.Date(DATE1),
```

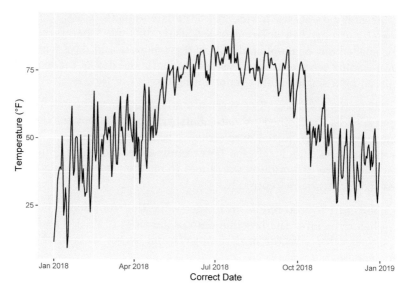

FIGURE 4.2
Time series plot with dates on the x-axis.

```
           DATE2B = as.Date(DATE2))
glimpse(datefile)
## Rows: 365
## Columns: 11
## $ YEAR   <dbl> 2018, 2018, 2018, 2018, 2018, 2018, 2018, 2~
## $ STID   <chr> "COOK", "COOK", "COOK", "COOK", "COOK", "CO~
## $ MONTH  <dbl> 1, 1, 1, 1, 1, 1, 1, 1, 1, 1, 1, 1, 1, 1, 1~
## $ DAY    <dbl> 1, 2, 3, 4, 5, 6, 7, 8, 9, 10, 11, 12, 13, ~
## $ TAVG   <dbl> 11.50, 16.86, 21.19, 24.56, 34.11, 36.92, 3~
## $ DATE   <date> 2018-01-01, 2018-01-02, 2018-01-03, 2018-0~
## $ DATE1  <chr> "2018-1-1", "2018-1-2", "2018-1-3", "2018-1~
## $ DATE2  <chr> "1-1-2018", "1-2-2018", "1-3-2018", "1-4-20~
## $ DATE3  <chr> "January 1, 2018", "January 2, 2018", "Janu~
## $ DATE1B <date> 2018-01-01, 2018-01-02, 2018-01-03, 2018-0~
## $ DATE2B <date> 0001-01-20, 0001-02-20, 0001-03-20, 0001-0~
```

The new DATE1B and DATE2B columns contain date objects. However, if we tried to convert the DATE3 column, R would return an error because as.Date() is not able to recognize the month names or the comma and space separators.

The conversion from the character in DATE1 to Date in DATE1B is successful, and a valid time series plot can be generated using DATE1B (Figure 4.3).

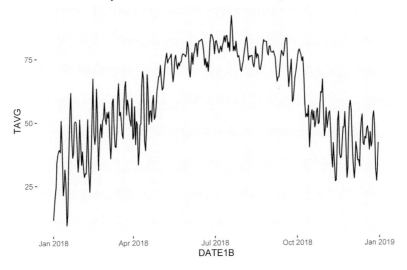

FIGURE 4.3

Time series plot with dates successfully converted using as.Date().

```
ggplot(data = datefile) +
  geom_line(aes(x = DATE1B, y = TAVG)) +
  ggtitle("DATE1 corrected converted to a date with as.Date()")
```

The conversion of DATE2, which is in MM-DD-YYYY format instead of YYYY-MM-DD format, produces a valid Date object in column DATE2B. However, by default, the as.Date() function assumes that the date string is in Year-Day-Month format, dates in the MM-DD-YYYY format are either read incorrectly or assigned a value of NA. The resulting time series graph shows dates from a range of years at the beginning of the Common Era (Figure 4.4). These are clearly the wrong dates for the 2018 weather dataset.

```
ggplot(data = datefile) +
  geom_line(aes(x = DATE2B, y = TAVG)) +
  ggtitle("DATE2 incorrectly converted to a date with as.Date()")
## Warning: Removed 221 row(s) containing missing values
## (geom_path).
```

To convert DATE2 to a date object using as.Date(), additional information about the formatting of the date string must be provided as a function argument. This can be done for specifying an additional string with formatting codes. An alternative is to use a package with functions that simplify the creation and

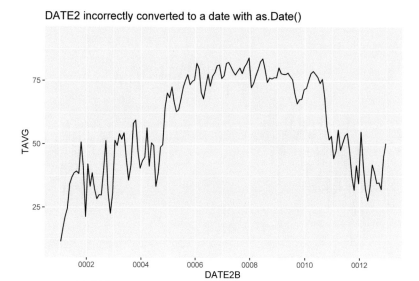

FIGURE 4.4
Time series plot with dates unsuccessfully converted using as.Date().

manipulation of Date objects. This book uses **lubridate**, another member of the **tidyverse** collection of packages.

In the following example, the ymd() function from **lubridate** is used to transform character strings in YYYY-MM-DD format into date objects, and the mdy() function is used to transform the MM-DD-YYYY format.

```
datefile <- datefile %>%
  mutate(DATE1C = ymd(DATE1),
         DATE2C = mdy(DATE2),
         DATE3C = mdy(DATE3))
glimpse(datefile)
## Rows: 365
## Columns: 14
## $ YEAR    <dbl> 2018, 2018, 2018, 2018, 2018, 2018, 2018, 2~
## $ STID    <chr> "COOK", "COOK", "COOK", "COOK", "COOK", "CO~
## $ MONTH   <dbl> 1, 1, 1, 1, 1, 1, 1, 1, 1, 1, 1, 1, 1, 1, 1~
## $ DAY     <dbl> 1, 2, 3, 4, 5, 6, 7, 8, 9, 10, 11, 12, 13, ~
## $ TAVG    <dbl> 11.50, 16.86, 21.19, 24.56, 34.11, 36.92, 3~
## $ DATE    <date> 2018-01-01, 2018-01-02, 2018-01-03, 2018-0~
## $ DATE1   <chr> "2018-1-1", "2018-1-2", "2018-1-3", "2018-1~
## $ DATE2   <chr> "1-1-2018", "1-2-2018", "1-3-2018", "1-4-20~
## $ DATE3   <chr> "January 1, 2018", "January 2, 2018", "Janu~
```

```
## $ DATE1B <date> 2018-01-01, 2018-01-02, 2018-01-03, 2018-0~
## $ DATE2B <date> 0001-01-20, 0001-02-20, 0001-03-20, 0001-0~
## $ DATE1C <date> 2018-01-01, 2018-01-02, 2018-01-03, 2018-0~
## $ DATE2C <date> 2018-01-01, 2018-01-02, 2018-01-03, 2018-0~
## $ DATE3C <date> 2018-01-01, 2018-01-02, 2018-01-03, 2018-0~
```

The DATE1C, DATE2C, and DATE3C columns all contain the correct dates now.

A handy feature of the **lubridate** functions is that they are very smart about interpreting in a variety of date formats without the user having to provide explicit formatting information. They can read dates parsed with dashes, slashes, or spaces, and they can deal with months and days with and without leading zeros. They can also deal with other types of date formats, such as the month names in the DATE3 column.

4.2 Other lubridate Operators and Functions

It is common to store different date elements, such as year, month, and day, in different database columns. This is a very effective way of representing date information in a database, since the date components can be stored as simple integers, and the meanings of the values in the year, month, and day columns are unambiguous.

The **lubridate** functions can be used to combine data from multiple columns into a single date object. A straightforward way to do this is to first use the paste() function to combine multiple values into a date string and then pass the string to the appropriate **lubridate** function. Here, the variables in the YEAR, MONTH, and DAY columns are combined to create the DAY4 column.

```
datefile <- datefile %>%
  mutate(DATE4 =ymd(paste(YEAR, MONTH, DAY, sep="-")))
glimpse(datefile)
## Rows: 365
## Columns: 15
## $ YEAR    <dbl> 2018, 2018, 2018, 2018, 2018, 2018, 2018, 2~
## $ STID    <chr> "COOK", "COOK", "COOK", "COOK", "COOK", "CO~
## $ MONTH   <dbl> 1, 1, 1, 1, 1, 1, 1, 1, 1, 1, 1, 1, 1, 1, 1~
## $ DAY     <dbl> 1, 2, 3, 4, 5, 6, 7, 8, 9, 10, 11, 12, 13, ~
## $ TAVG    <dbl> 11.50, 16.86, 21.19, 24.56, 34.11, 36.92, 3~
## $ DATE    <date> 2018-01-01, 2018-01-02, 2018-01-03, 2018-0~
## $ DATE1   <chr> "2018-1-1", "2018-1-2", "2018-1-3", "2018-1~
```

```
## $ DATE2  <chr> "1-1-2018", "1-2-2018", "1-3-2018", "1-4-20~
## $ DATE3  <chr> "January 1, 2018", "January 2, 2018", "Janu~
## $ DATE1B <date> 2018-01-01, 2018-01-02, 2018-01-03, 2018-0~
## $ DATE2B <date> 0001-01-20, 0001-02-20, 0001-03-20, 0001-0~
## $ DATE1C <date> 2018-01-01, 2018-01-02, 2018-01-03, 2018-0~
## $ DATE2C <date> 2018-01-01, 2018-01-02, 2018-01-03, 2018-0~
## $ DATE3C <date> 2018-01-01, 2018-01-02, 2018-01-03, 2018-0~
## $ DATE4  <date> 2018-01-01, 2018-01-02, 2018-01-03, 2018-0~
```

Other functions in the **lubridate** package can be used to extract information from Date objects and to do calculations on them. The following example extracts two dates from the DATE column and stores them as separate objects.

```
date1 <- datefile$DATE[1]
date2 <- datefile$DATE[180]
date1
## [1] "2018-01-01"
date2
## [1] "2018-06-29"
```

The examples extract the month, day of the month, year, week of the year, day of the week, and day of the year associated with a Date object. They can be helpful for converting a date object into numbers for storage in separate columns. The day of the year is an important variable when analyzing seasonal patterns of environmental change.

```
month(date2)
## [1] 6
day(date2)
## [1] 29
year(date2)
## [1] 2018
week(date2)
## [1] 26
wday(date2)
## [1] 6
yday(date2)
## [1] 180
```

Operators can also be used with date objects. In particular, the - operator is useful for determining the time interval between two dates.

```
date2 - date1
## Time difference of 179 days
date1 - date2
## Time difference of -179 days
```

4.3   Practice

Use the code provided below to create two vectors: `datetxt` contains dates stored as text in month-day-year format, and `relhumid` contains daily measurements of minimum relative humidity. Convert `datetxt` to a date object and generate a line graph of changes in relative humidity over time. Create two new vectors that contain the day of the year and the day of the week associated with each observation.

```
datetxt <- c("3-13-2017", "3-14-2017", "3-15-2017", "3-16-2017",
             "3-17-2017", "3-18-2017", "3-19-2017", "3-20-2017",
             "3-21-2017", "3-22-2017")
relhumid <- c(56, 59, 49, 51, 53, 58, 65, 62, 66, 69)
```

5

Vector Geospatial Data

Until now, this book has focused on non-spatial tabular datasets. This chapter will make the shift to processing and visualizing vector geospatial data. Vector data consists of *features* that represent the geographic phenomenon. Points can be used to represent small objects such as wells, weather stations, or field plot locations. Lines represent one-dimensional linear features such as roads and streams. Polygons represent two-dimensional natural features such as lakes, vegetation patches, and burn scars as well as administrative units such as counties, districts, and counties. The type of geometry used to represent a particular feature may depend on the scale of the measurement. Points can be used to represent city locations at a continental scale, whereas a polygon would be used to map the boundary of an individual city.

Each feature in a vector dataset is described by a set of spatial coordinates and one or more attributes that characterize the feature. For example, a vector dataset of forest stands would include the vertices of each stand boundary polygon as well an attribute table containing variables such as stand age, tree species abundances, and timber volume for each stand. This book uses the **sf** package (Pebesma, 2022) to work with vector datasets as "simple features" in R. This approach uses data frame objects to store vector data. Each feature is represented by one row in the data frame, with attributes stored as columns and spatial information stored in a special geometry column. This means that we can continue to use the same **dplyr** and **tidyr** functions that have already been covered to work with geospatial data. The **ggplot2** package can also be used to generate maps from data frames containing geospatial data.

This chapter will use the **sf** package along with several of the **tidyverse** packages from the previous chapters and several other packages that contain some helpful graphics functions. If these packages are not already installed, you can install them using the `install.packages()` function or by selecting Tools > Install Packages in the RStudio menu.

```
library(sf)
library(rgdal)
library(ggplot2)
library(dplyr)
library(tidyr)
```

DOI: 10.1201/9781003326199-5

```
library(scales)
library(RColorBrewer)
library(units)
library(cowplot)
```

5.1 Importing Geospatial Data

This chapter uses several datasets that characterize the historical distribution
of tornadoes in the state of Oklahoma. They are provided in ESRI shapefile
format, which was developed several decades ago but remains one of the widely
used file formats for vector geospatial data. It is a multiple file format, where
separate files contain the feature geometries, attribute table, spatial indices,
and coordinate reference system. All files have the same filename with different
extensions. The shapefiles are imported to sf objects using the st_read()
function. The quiet = FALSE argument suppresses output to the console when
importing spatial datasets. To read in a shapefile, it is only necessary to specify
the filename with a ".shp" extension. However, all the files, including the ".shp"
file as well as the ".dbf", ".shx", and ".prj" files, need to be present in the
directory from which the data are read. Remember that all of the code in this
book is written with the assumption that data are being read from the user's
working directory.

The ok_counties.shp dataset contains county boundaries for the state of
Oklahoma. The ok_tornado_point.shp dataset and the ok_tornado_path.shp
dataset contain historical information about tornadoes in Oklahoma. The
points are the initial locations of tornado touchdown, and the paths are lines
that identify the path of each tornado after touchdown. These data were derived
from larger, national-level datasets generated by the National Oceanographic
and Atmospheric Administration (NOAA) National Weather Service Storm
Prediction Center (https://www.spc.noaa.gov/gis/svrgis/).

```
okcounty <- st_read("ok_counties.shp", quiet = TRUE)
tpoint <- st_read("ok_tornado_point.shp", quiet = TRUE)
tpath <- st_read("ok_tornado_path.shp", quiet = TRUE)
class(okcounty)
## [1] "sf"          "data.frame"
glimpse(okcounty)
## Rows: 77
## Columns: 8
## $ STATEFP   <chr> "40", "40", "40", "40", "40", "40", "40",~
```

```
## $ COUNTYFP <chr> "077", "025", "011", "107", "105", "153",~
## $ COUNTYNS <chr> "01101826", "01101800", "01101793", "0110~
## $ AFFGEOID <chr> "0500000US40077", "0500000US40025", "0500~
## $ GEOID    <chr> "40077", "40025", "40011", "40107", "4010~
## $ NAME     <chr> "Latimer", "Cimarron", "Blaine", "Okfuske~
## $ LSAD     <chr> "06", "06", "06", "06", "06", "06", "06",~
## $ geometry <POLYGON [°]> POLYGON ((-95.50766 35.0292..., P~
```

The sf objects contain a column called geometry. This is a special column that contains the geospatial information about the location of each feature. This column should not be modified directly. It is used by the functions in the **sf** package for geospatial data processing.

An older package that is still frequently used for working with geospatial data in R is **sp**. This package has been around for much longer than **sf**, and there are a number of other R packages that work with geospatial data objects defined in **sp**. Therefore, it is sometimes necessary to convert geospatial data between these two object classes. The following examples show how to convert okcounty into an **sp** SpatialPolygonsDataFrame object and then back to an sf object.

```
okcounty_sp <- as(okcounty, 'Spatial')
class(okcounty_sp)
## [1] "SpatialPolygonsDataFrame"
## attr(,"package")
## [1] "sp"
okcounty_sf <- st_as_sf(okcounty_sp)
class(okcounty_sf)
## [1] "sf"          "data.frame"
```

5.2 Creating Simple Maps

To generate a map of counties using ggplot() with a **sf** object, the geom_sf() function is used. This map shows the borders of all counties in Oklahoma. No plot aesthetics are specified, and fill = NA makes the counties transparent. The map is displayed over the default gray background, and graticules of longitude and latitude are shown and labeled. (Figure 5.1).

```
ggplot(data = okcounty) +
  geom_sf(fill = NA)
```

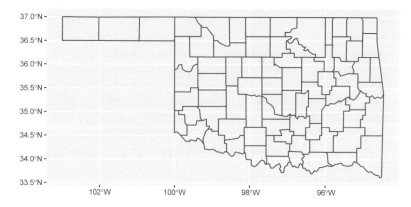

FIGURE 5.1
Oklahoma county boundaries.

The tornado dataset contains data from 1950–2021. If the objective is to map tornadoes for only recent years, then the sf object needs to be modified to contain only the desired years. Because sf objects are a type of data frame, they can be modified using the **dplyr** functions that were covered in Chapter 3. The tornado datasets contain a number of columns with information on the timing, locations, and effects of the tornadoes.

```
names(tpoint)
##  [1] "om"      "yr"      "mo"      "dy"      "date"
##  [6] "time"    "tz"      "st"      "stf"     "stn"
## [11] "mag"     "inj"     "fat"     "loss"    "closs"
## [16] "slat"    "slon"    "elat"    "elon"    "len"
## [21] "wid"     "fc"      "geometry"
```

The yr column indicates the year of each tornado and can be used to filter the data to a smaller year range from 2016–2021. This column is retained in the output along with om, a unique ID code, and date, the date of each tornado.

```
tpoint_16_21 <- tpoint %>%
  filter(yr >= 2016 & yr <= 2021) %>%
  select(om, yr, date)
tpath_16_21 <- tpath %>%
  filter(yr >= 2016 & yr <= 2021) %>%
  select(om, yr, date)
```

Now, the tornado points and paths can be mapped. Two geom_sf() functions are used, one to map the county boundaries and another to map the tornadoes.

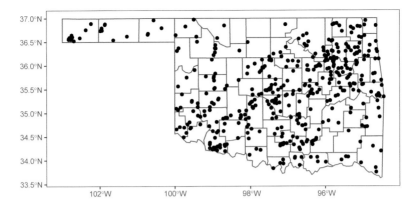

FIGURE 5.2
Initiation points of tornadoes in Oklahoma from 2016–2021.

Because each function maps a different dataset, the data argument must be provided in each `geom_sf()` function instead of in the `ggplot()` function. The `theme_bw()` function is also used to display the map over a white background while retaining the graticules (Figure 5.2).

```
ggplot() +
  geom_sf(data = okcounty, fill = NA) +
  geom_sf(data = tpoint_16_21) +
  theme_bw()
```

To make the tornado path lines easier to see in relation to the county boundaries, they are displayed in red and their sizes are increased to be larger than the default line width of 0.5 (Figure 5.3).

```
ggplot() +
  geom_sf(data = okcounty, fill = NA) +
  geom_sf(data = tpath_16_21,
          color = "red",
          size = 1) +
  theme_bw()
```

To view the years of the tornadoes on the map, an aesthetic can be specified. The `color` argument is specified as `as.factor(yr)` so that the year is displayed as a discrete variable instead of a continuous variable. The `scale_color_discrete()` function is then used to set the legend name. The `theme_void()` function removes the plot axes and labels and shows only the map (Figure 5.4).

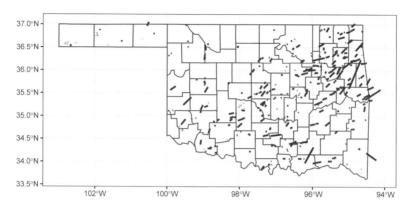

FIGURE 5.3
Paths of tornadoes in Oklahoma from 2016–2021.

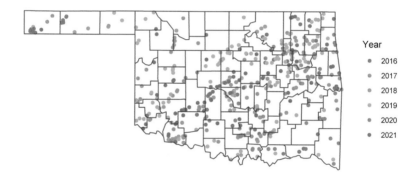

FIGURE 5.4
Initiation points of tornadoes in Oklahoma from 2016–2021 with years represented by the color aesthetic.

```
ggplot() +
  geom_sf(data = tpoint_16_21,
          aes(color = as.factor(yr))) +
  geom_sf(data = okcounty, fill = NA) +
  scale_color_discrete(name = "Year") +
  coord_sf(datum = NA) +
  theme_void()
```

Alternately, `facet_wrap()` can be used to display the tornadoes for each year on a separate map. The sizes of the points are reduced slightly so that they are easier to see on the smaller maps (Figure 5.5).

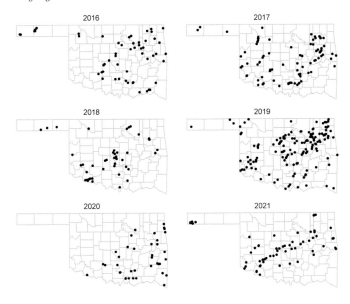

FIGURE 5.5
Initiation points of tornadoes in Oklahoma from 2016–2021 with years mapped as separate facets.

```
ggplot() +
  geom_sf(data = okcounty,
          fill = NA,
          color = "gray") +
  geom_sf(data = tpoint_16_21, size = 0.75) +
  facet_wrap(vars(yr), ncol = 2) +
  coord_sf(datum = NA) +
  theme_void()
```

5.3 Overlaying Vector Datasets

The number of tornado points in each county can be calculated using the st_join() function to carry out a spatial join. This is different from the join functions covered in Chapter 3 in that st_join() links rows from the two tables based on the spatial locations instead of their attributes. Afterward, each row in countypnt contains additional columns from the okcounty dataset that correspond to the county that the tornado is within.

```
countypnt <- st_join(tpoint_16_21, okcounty)
glimpse(countypnt)
## Rows: 434
## Columns: 11
## $ om       <dbl> 613662, 613675, 613676, 613677, 613678, 6~
## $ yr       <dbl> 2016, 2016, 2016, 2016, 2016, 2016, 2016,~
## $ date     <chr> "2016-03-23", "2016-03-30", "2016-03-30",~
## $ STATEFP  <chr> "40", "40", "40", "40", "40", "40", "40",~
## $ COUNTYFP <chr> "001", "113", "105", "131", "035", "139",~
## $ COUNTYNS <chr> "01101788", "01101844", "01101840", "0110~
## $ AFFGEOID <chr> "0500000US40001", "0500000US40113", "0500~
## $ GEOID    <chr> "40001", "40113", "40105", "40131", "4003~
## $ NAME     <chr> "Adair", "Osage", "Nowata", "Rogers", "Cr~
## $ LSAD     <chr> "06", "06", "06", "06", "06", "06", "06",~
## $ geometry <POINT [°]> POINT (-94.5042 35.6916), POINT (-9~
```

The joined data frame still contains one record for each tornado. To compute
the total number of tornadoes per county, countypnt must be grouped by the
GEOID county code. Then, the number of tornadoes in each county can be
calculated using summarize() with the n() function to count the records in
each group. Before grouping and summarizing, countypnt must be converted
from an sf object to a normal data frame using st_drop_geometry().

```
countypnt <- st_drop_geometry(countypnt)
countysum <- countypnt %>%
  group_by(GEOID) %>%
  summarize(tcnt = n())
glimpse(countysum)
## Rows: 75
## Columns: 2
## $ GEOID <chr> "40001", "40005", "40007", "40009", "40011",~
## $ tcnt  <int> 6, 3, 4, 8, 1, 4, 10, 5, 7, 5, 3, 12, 10, 5,~
```

Next, okcounty is joined to countysum so that each polygon is associated with
the appropriate tornado summary. Using left_join() instead of inner_join()
ensures that all of the county polygons are retained in the output of the join.
A few counties that had no tornadoes during 2016–2021 are missing from
countysum, and therefore have NA values in the joined table. In this case, we
know that NA means zero tornadoes, so the NA values are replaced by zeroes.
The mutate() function computes the area of each county using st_area() and
then calculates the density of tornadoes per county. Density is initially in
tornadoes per square meter but is converted to tornadoes per 1000 km². The
st_area() function returns a column with explicit measurement units, but
these are removed using the drop_units() function for simplicity. For more

information about measurement units, consult the documentation of the **units** package (Pebesma et al., 2022).

```
countymap <- okcounty %>%
  left_join(countysum, by = "GEOID") %>%
  replace(is.na(.), 0) %>%
  mutate(area = st_area(okcounty),
         tdens = 10^6 * 10^3 * tcnt / area) %>%
  drop_units()
glimpse(countymap)
## Rows: 77
## Columns: 11
## $ STATEFP  <chr> "40", "40", "40", "40", "40", "40", "40",~
## $ COUNTYFP <chr> "077", "025", "011", "107", "105", "153",~
## $ COUNTYNS <chr> "01101826", "01101800", "01101793", "0110~
## $ AFFGEOID <chr> "0500000US40077", "0500000US40025", "0500~
## $ GEOID    <chr> "40077", "40025", "40011", "40107", "4010~
## $ NAME     <chr> "Latimer", "Cimarron", "Blaine", "Okfuske~
## $ LSAD     <chr> "06", "06", "06", "06", "06", "06", "06",~
## $ tcnt     <dbl> 1, 12, 1, 10, 6, 2, 6, 0, 4, 9, 3, 10, 12~
## $ geometry <POLYGON [°]> POLYGON ((-95.50766 35.0292..., P~
## $ area     <dbl> 1890663261, 4766283042, 2427121029, 16572~
## $ tdens    <dbl> 0.5289149, 2.5176851, 0.4120108, 6.034094~
```

The `st_write()` function can be used to save `sf` objects to a variety of file formats. In many cases, the function can determine the output format from the output filename extension. The following code saves the county-level tornado summaries in ESRI shapefile format. The `append = FALSE` option overwrites the shapefile if it already exists.

```
st_write(countymap,
         dsn = "oktornadosum.shp",
         append = FALSE)
```

An alternative format for storing geospatial data is the Open Geospatial Consortium (OGC) GeoPackage. The data are stored in an SQLite database that may contain one or more layers. In this example, the `delete_dsn = TRUE` argument overwrites the entire GeoPackage. Leaving this argument at its default value of `FALSE` would add a new layer to an existing database.

```
st_write(countymap,
         dsn = "oktornado.gpkg",
         layer = "countysum",
         delete_dsn = TRUE)
```

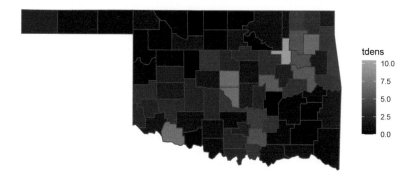

FIGURE 5.6
Densities of tornadoes in Oklahoma counties from 2016–2021 mapped as a
choropleth.

Another commonly-used open geospatial data format is GeoJSON. It is based
on Javascript Object Notation (JSON), a human-readable text format that
stores data in ASCII files. The layer_options argument must be set to "RFC7946
= YES" to save the data in the newest GeoJSON specification.

```
st_write(countymap, "oktornado.geojson",
         layer_options = "RFC7946 = YES")
```

5.4   Choropleth Maps

Another way to display the tornadoes is as a choropleth map, where summary
statistics for each county are displayed as different colors. The county-level
tornado density can be as a choropleth using the fill aesthetic with geom_sf().
By default, the fill colors are based on a dark-to-light blue color ramp. The
theme_void() function eliminates the axes and graticules and displays only the
map on a white background (Figure 5.6).

```
ggplot(data = countymap) +
  geom_sf(aes(fill = tdens)) +
  theme_void()
```

In general, choropleth maps are most effective when mapping rates, such as
the density of tornadoes per area or the prevalence of a disease per number
of people. In some cases, it is necessary to map count data such as the
number of tornadoes or disease cases in each county. People tend to naturally

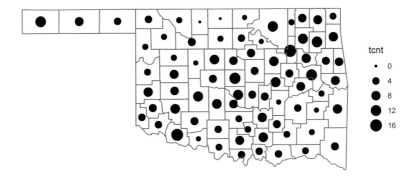

FIGURE 5.7
Numbers of tornadoes in Oklahoma counties from 2016–2021 mapped as graduated symbols.

associate sizes with quantities when viewing maps, and displaying the counts as graduated symbols is often an effective approach.

To map symbols, the county polygons must first be converted to points. The st_centroid() generates a point feature located at the centroid of each county. The st_geometry_type() function returns the feature geometry type. Setting by_geometry = FALSE returns one geometry type for the entire dataset instead of for every feature.

```
st_geometry_type(okcounty, by_geometry=FALSE)
## [1] POLYGON
## 18 Levels: GEOMETRY POINT LINESTRING POLYGON ... TRIANGLE
okcntrd = st_centroid(countymap)
st_geometry_type(okcntrd, by_geometry = FALSE)
## [1] POINT
## 18 Levels: GEOMETRY POINT LINESTRING POLYGON ... TRIANGLE
```

The tornado counts can be mapped using the okcentrd dataset with the size aesthetic. One point is plotted for each county centroid, and the size of the point is proportional to the number of tornadoes in the county (Figure 5.7).

```
ggplot() +
  geom_sf(data = okcntrd, aes(size = tcnt)) +
  geom_sf(data = okcounty, fill = NA) +
  theme_void()
```

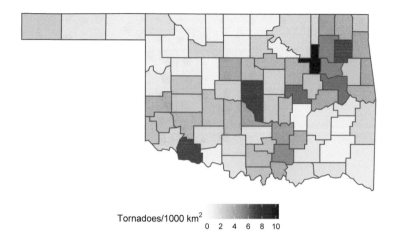

FIGURE 5.8
Densities of tornadoes in Oklahoma counties from 2016–2021 mapped as a choropleth with a custom palette.

5.5   Modifying the Appearance of the Map

Additional **ggplot2** functions can be added to improve the appearance of the map. The `scale_fill_distiller()` function allows the specification of a different color ramp. This example uses the "YlOrRd" palette from the **RColorBrewer** package. The `pretty_breaks()` function from the **scales** package is used to automatically select a set of breaks for the legend. The legend is moved to the bottom of the map to better accommodate the longer legend title (Figure 5.8).

Note that using a "complete" theme like `theme_void()` will remove any settings made by a previous `theme()` function. Therefore, it is necessary to call `theme_void()` before `theme()` in the `ggplot()` call to implement the combined settings. Also, note how the `expression()` function is used in specifying the name argument for `scale_fill_distiller()` to combine text with a superscript.

```
ggplot(data = countymap) +
  geom_sf(aes(fill = tdens)) +
  scale_fill_distiller(name = expression("Tornadoes/1000 km"^2),
                       palette = "YlOrRd",
                       breaks = pretty_breaks(),
                       direction = 1) +
  theme_void() +
  theme(legend.position = "bottom")
```

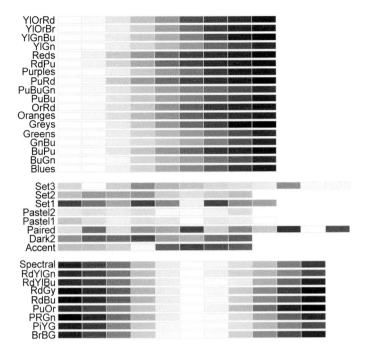

FIGURE 5.9
All the palettes available in the RColorBrewer package.

The **RColorBrewer** package provides a selection of palettes designed for choropleth mapping (Harrower and Brewer, 2003). The `display_brewer_all()` function generates a chart with an example of all the available palettes (Figure 5.9).

```
display.brewer.all()
```

There are three types of ColorBrewer palettes. The top group is Figure 5.9 contains sequential palettes that are suitable for mapping ordered data along numerical scales (e.g., temperatures ranging from 0 to 30 degrees C) or ordinal categories (e.g., temperatures classified as cold, warm, and hot). These sequential palettes may include a single color or multiple colors but have no clear breaks in the scale. The middle group in Figure 5.9 contains qualitative palettes, which use color to distinguish between different categories without implying order. The lower group in Figure 5.9 contains divergent palettes that emphasize a specific breakpoint in the data. Divergent palettes are often used to indicate values that are above or below the mean or to highlight values that are higher or lower than zero. More details about these palettes, including recommendations for color schemes that are most effective for different types

YlGnBu (sequential)

FIGURE 5.10
The ColorBrewer yellow-green-blue color palette with five categories.

of computer displays and for accommodating color-blind viewers, are available
at http://colorbrewer2.org.

The ColorBrewer palettes each contain a finite number of colors that are
intended to be associated with categories in a choropleth map. Note that the
scale_fill_distiller() function used to generate the color scale for the map
in Figure 5.8 operates a bit differently. This function takes a ColorBrewer
palette and converts it to a continuous color ramp. The next map example
will show how to define categories and map each one as a distinctive color. To
view the colors for a given number of categories and a specific palette, the
display.brewer.pal() function is used with the number of categories as the
first argument and the palette name as the second palette (Figure 5.10).

```
display.brewer.pal(5, "YlGnBu")
```

Rather than using continuous scales for color and size, it is often recommended
to aggregate the data into a small number of classes (typically 3-6). Using
discrete classes makes it easier to associate each color or symbol in the map
with a specific range of values. To accomplish this step, we need to add a couple
of new classified variables using mutate(). The cut() function is used to split
the continuous variables based on user-specified breaks. The incidence variable
is split based on quantiles (i.e., percentiles) defined in the qbrks object. The
population breaks are manually specified.

```
numclas <- 4
qbrks <- seq(0, 1, length.out = numclas + 1)
```

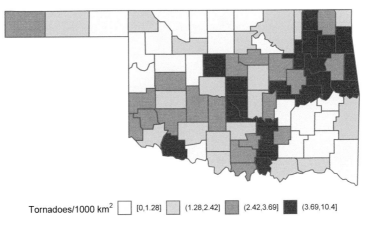

FIGURE 5.11
Densities of tornadoes in Oklahoma counties from 2016–2021 mapped as a choropleth with four categories.

```
qbrks
## [1] 0.00 0.25 0.50 0.75 1.00
countymap <- countymap %>%
  mutate(tdens_c1 = cut(tdens,
                        breaks = quantile(tdens, breaks = qbrks),
                        include.lowest = T))
```

The new `tdens_c1` column is a discrete factor instead of a continuous numerical variable as in the previous example. As a result, the `scale_fill_brewer()` function is now used instead of `scale_fill_distiller()`. The comma-separated numbers specify the range of tornado densities for each of the categories (Figure 5.11).

```
ggplot(data = countymap) +
  geom_sf(aes(fill = tdens_c1)) +
  scale_fill_brewer(name = expression("Tornadoes/1000 km"^2),
                    palette = "YlOrRd") +
  theme_void() +
  theme(legend.position = "bottom")
```

Similar to choropleth maps, graduated symbol maps are often easier to interpret if they include a limited number of symbol sizes. To generate a classified map of tornado counts, they are converted to discrete categories using the `cut()` function. Instead of using quantiles, the breakpoints for the classification are selected manually and stored in the `brkpts` vector.

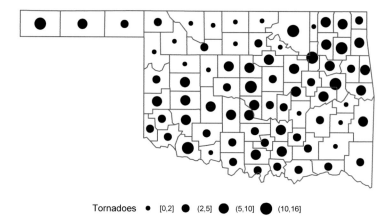

FIGURE 5.12

Numbers of tornadoes in Oklahoma counties from 2016–2021 mapped as graduate symbols with categories.

```
maxcnt <- max(okcntrd$tcnt)
brkpts <- c(0, 2, 5, 10, maxcnt)
okcntrd <- okcntrd %>%
  mutate(tcnt_cl = cut(tcnt,
                       breaks = brkpts,
                       include.lowest = T))
```

The resulting map has four symbol sizes, each associated with a specific range of tornado counts (Figure 5.12).

```
ggplot(data = okcntrd) +
  geom_sf(aes(size = tcnt_cl)) +
  scale_size_discrete(name="Tornadoes") +
  geom_sf(data = okcounty, fill = NA) +
  theme_void() +
  theme(legend.position = "bottom")
```

5.6   Exporting Graphics Output

By default, maps and charts generated using `ggplot()` are output to the Plots tab in the lower right-hand corner of the RStudio GUI. However, it is often necessary to export maps and other graphics to files and explicitly specify their dimensions and resolution. This is usually the case when generating graphics for publications that must meet specific size and formatting criteria. The simplest way to do this is with the ggsave() function. This example exports the most recent output of ggplot() to a portable network graphics (PNG) file called tornado.png with dimensions of 6 x 4 inches and a resolution of 300 pixels per inch. Other units besides inches can be used by specifying the units argument.

```r
ggsave("tornado.png",
       width = 6,
       height = 4,
       dpi = 300)
ggsave("tornado2.png",
       width = 15,
       height = 10,
       units = "cm",
       dpi = 100)
```

A variety of graphical file types are supported, including TIFF, JPEG, and PDF files. Note that the arguments vary with the type of file being created. When saving a JPEG, the quality option can be specified. When saving a TIFF, a compression type can be specified. When saving a PDF, the dpi argument is ignored.

```r
ggsave("tornado.jpeg",
       width = 6,
       height = 4,
       dpi = 300,
       quality = 90)
ggsave("tornado.tiff",
       width = 6,
       height = 4,
       dpi = 300,
       compression = "lzw")
ggsave("tornado.pdf",
       width = 6,
       height = 4)
```

The output of ggplot() can also be saved as an R object that can be output to graphics files using ggsave(). The plot argument is used to specify the ggplot object to be saved.

```r
choropleth <- ggplot(data = countymap) +
  geom_sf(aes(fill = tdens_c1)) +
```

```
    scale_fill_brewer(name="Density",
                       palette = "YlOrRd") +
    theme_void()

gradsymbol <- ggplot(data = okcntrd) +
    geom_sf(aes(size = tcnt_c1)) +
    scale_size_discrete(name="Count") +
    geom_sf(data = okcounty, fill = NA) +
    theme_void()

ggsave("choropleth.tiff",
       plot = choropleth,
       width = 6,
       height = 4,
       dpi = 300,
       compression = "lzw")

ggsave("gradsymbol.tiff",
       plot = gradsymbol,
       width = 6,
       height = 4,
       dpi = 300,
       compression = "lzw")
```

Saved graphs and maps can also be combined into a composite figure using the **cowplot** package (Wilke, 2020). The `plot_grid()` function provides a variety of options for arranging figures in a regular grid. This basic example provides a label for each subplot and specifies additional arguments to plot the maps in a single column, justify the labels, move the labels to the top of each map, and align the maps horizontally and vertically, so they are exactly the same size.

```
plot_grid(choropleth, gradsymbol,
          labels = c("A) Choropleth Map",
                     "B) Graduated Symbol Map",
                     label_size = 12),
          ncol = 1,
          hjust = 0,
          label_x = 0,
          align = "hv")
```

A variety of additional resources for working with the **sf** package can be found at `https://r-spatial.github.io/sf/index.html`. These include a link to the **sf** "cheatsheet" as well as a variety of articles, vignettes, and blog posts that provide additional examples of how to work with vector geospatial data using

A) Choropleth Map

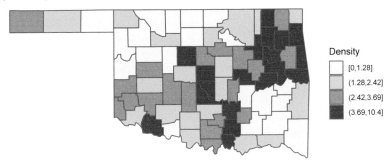

B) Graduated Symbol Map

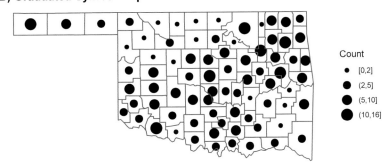

FIGURE 5.13
Multiple map figures generated with the cowplot package.

this package. The book *Geocomputation with R* by Robin Lovelace and others
(Lovelace et al., 2019) is also an excellent reference that encompasses **sf** as well
as many other R packages for working with geospatial data. Look for their
forthcoming second edition for the most up-to-date information on geospatial
data processing with R.

5.7 Practice

1. Generate a map of tornado paths where the paths from each year
   are displayed as a different color, similar to the map of tornado
   points in Figure 5.4. Create a composite figure containing the map
   of tornado paths and the map of tornado points using plot_grid().

2. Summarize the density of tornado points by both county and year and generate a faceted plot that displays maps of county-level tornado density from 2016–2021.

3. Generate four choropleth maps of tornado density based on quantile breaks with numbers of classes ranging from 3 to 6. Create a composite figure containing the four maps using `plot_grid()` and examine how the number of classes changes the interpretation of the map.

6

Raster Geospatial Data—Continuous

Raster data are fundamentally different from vector data in that they are referenced to a regular grid of rectangular (usually square) cells. The spatial characteristics of a raster dataset are defined by its spatial resolution (the height and width of each cell) and its origin (typically the upper left corner of the raster grid, which is associated with a location in a coordinate reference system). The raster data format has several advantages and limitations compared to vector data. Continuous variables such as elevation, temperature, and precipitation as well as categorical data such as discrete vegetation types and land cover classes can often be stored and manipulated more efficiently as rasters. However, the raster format can be imprecise and inefficient for point, line, and small polygon features such as plot locations, roads, streams, and water bodies.

The **terra** package provides a variety of specialized classes and functions for importing, processing, analyzing, and visualizing raster datasets (Hijmans, 2022). It is intended to replace the **raster** package, which has been the main raster data package in R for many years. The data objects and the function syntax in the two packages are very similar, and longtime **raster** users should find it straightforward to work with **terra**. However, the **terra** package contains several major improvements, including faster processing speed for large rasters. To avoid unnecessary confusion, the code in this book is based entirely on the newer **terra** package. There are several other packages that can handle raster data. For example, the older **sp** package supports the `SpatialGridDataFrame` class for gridded datasets and **spatstat** supports the `im` class for pixel image objects. However, **terra** is by far the most flexible and powerful package currently available in R for handling raster data.

```
library(sf)
library(terra)
library(ggplot2)
```

6.1   Importing Raster Data

A raster object can be created by calling the `rast()` function and specifying an external image file as an argument. In this example, a dataset of land surface temperature (LST) measured by the MODIS sensor on board the Terra satellite is imported into R as a raster object. MODIS data, along with many other types of satellite remote sensing products, can be obtained from the United State Geological Survey's Land Processes Distributed Active Archive Center (`https://lpdaac.usgs.gov/`).

The resulting object belongs to the `SpatRaster` class. Invoking the `print()` function for a raster object provides information about the dimensions of the grid, cell size, geographic location, and other details.

```
lst <- rast("MOD11A2_2017-07-12.LST_Day_1km.tif")
class(lst)
## [1] "SpatRaster"
## attr(,"package")
## [1] "terra"
lst
## class       : SpatRaster
## dimensions  : 1110, 3902, 1  (nrow, ncol, nlyr)
## resolution  : 0.009009009, 0.009009009  (x, y)
## extent      : -104.4326, -69.27943, 30, 40  (xmin, xmax, ymin, ymax)
## coord. ref. : lon/lat WGS 84 (EPSG:4326)
## source      : MOD11A2_2017-07-12.LST_Day_1km.tif
## name        : MOD11A2_2017-07-12.LST_Day_1km
```

The `summary()` function provides information about the statistical distribution of raster values, with the `size` argument specifying the number of randomly sampled pixels to include in the summary.

```
summary(lst, size = 1e6)
##   MOD11A2_2017.07.12.LST_Day_1km
##   Min.   :    0
##   1st Qu.:    0
##   Median :14981
##   Mean   : 8321
##   3rd Qu.:15148
##   Max.   :16396
```

There are a number of helper functions, shown below, that can be used to extract specific characteristics of a `SpatRaster` object. The `ncell()`, `nrow()`, ncol(), and nlyr() functions return the numbers of grid cells, grid rows, grid columns, and raster layers in the dataset. The res() function returns the grid cell size.

```
ncell(lst)
## [1] 4331220
nrow(lst)
## [1] 1110
ncol(lst)
## [1] 3902
nlyr(lst)
## [1] 1
res(lst)
## [1] 0.009009009 0.009009009
```

The ext() function returns a SpatExtent object that contains the geographic coordinates of the raster extent. SpatExtent objects can be used to specify the extent of new raster objects or to crop rasters to a new extent.

```
lst_ext <- ext(lst)
class(lst_ext)
## [1] "SpatExtent"
## attr(,"package")
## [1] "terra"
lst_ext[1:4]
##        xmin       xmax       ymin       ymax
## -104.43258  -69.27943   30.00000   40.00000
```

The crs() function returns a character string containing details about the coordinate reference system (CRS) of the raster dataset. By default, the output is in well-known text (WKT) format. The parse = TRUE argument parses the output into a vector that is easier to read when printed to the R console.

```
crs(lst, parse = TRUE)
##  [1] "GEOGCRS[\"WGS 84\","
##  [2] "    DATUM[\"World Geodetic System 1984\","
##  [3] "        ELLIPSOID[\"WGS 84\",6378137,298.257223563,"
##  [4] "            LENGTHUNIT[\"metre\",1]]],"
##  [5] "    PRIMEM[\"Greenwich\",0,"
##  [6] "        ANGLEUNIT[\"degree\",0.0174532925199433]],"
##  [7] "    CS[ellipsoidal,2],"
##  [8] "        AXIS[\"geodetic latitude (Lat)\",north,"
##  [9] "            ORDER[1],"
## [10] "            ANGLEUNIT[\"degree\",0.0174532925199433]],"
```

```
## [11] "          AXIS[\"geodetic longitude (Lon)\",east,"
## [12] "              ORDER[2],"
## [13] "              ANGLEUNIT[\"degree\",0.0174532925199433]],"
## [14] "    ID[\"EPSG\",4326]]"
```

The describe = TRUE argument returns an abbreviated summary that includes the CRS name and European Petroleum Specialty Group (EPSG) code for the CRS. More details on working with coordinate reference systems are provided in Chapter 8.

```
crs(lst, describe = TRUE)
##     name authority code area            extent
## 1 WGS 84      EPSG 4326 <NA> NA, NA, NA, NA
```

Each layer in a SpatRaster object has a name. By default, this name is the same as the data file that was used to create the raster. Because file names are often long and difficult to interpret, it is useful to specify more readable layer names. Layer names can be extracted and assigned using the names() function.

```
names(lst)
## [1] "MOD11A2_2017-07-12.LST_Day_1km"
names(lst) <- c("temperature")
names(lst)
## [1] "temperature"
```

SpatRaster objects can be modified with various functions and used as input to mathematical expressions. In the LST dataset, missing data are coded as zeroes, and the raw digital numbers must be modified by a scaling factor of 0.02 to convert them into degrees Kelvin. The code below uses the ifel() function, which is analogous to the ifelse() function in base R, to replace zero values in the LST dataset with NA values. The first argument to this function is a logical expression, the second is the value to return where the expression is TRUE, and the third is the value to return where the expression is FALSE. The next expression applies the scaling factor to obtain degrees Kelvin and subtracts a constant to further convert the temperatures from Kelvin to Celsius. In both cases, the calculations are automatically carried out for every grid cell in the raster dataset.

```
lst <- ifel(lst == 0, NA, lst)
lst_c <- lst * 0.02 - 273.15
summary(lst_c, size = 1e6)
##     temperature
## Min.    : 9.1
```

```
##   1st Qu.:28.0
##   Median :29.6
##   Mean   :30.3
##   3rd Qu.:31.5
##   Max.   :54.8
##   NA's   :452091
```

The `global()` function can be used to generate statistical summaries of the pixels in a raster dataset. As demonstrated in Chapter 1, the `na.rm` argument must be TRUE to ignore missing data in the calculations.

```
global(lst_c, fun = "mean", na.rm=T)
##                   mean
## temperature 30.28687
global(lst_c, fun = "min", na.rm=T)
##                 min
## temperature 1.67
global(lst_c, fun = "max", na.rm=T)
##                  max
## temperature 54.77
global(lst_c, fun = "sd", na.rm=T)
##                   sd
## temperature 3.652403
```

Raster objects can be exported using the `writeRaster()` function. The format of the exported image is specified using the `format` argument. Common output formats are `GTiff` (GeoTiff), `ascii` (ESRI ASCII text), `CDF` (NetCDF), and `HFA` (ERDAS Imagine). The `overwrite=TRUE` argument will replace existing files with the same name.

```
writeRaster(lst_c,
            filename = "MOD11A2_2017-07-12.LST_Day_1km_DegC.tif",
            filetype="GTiff", overwrite=TRUE )
```

6.2   Maps of Raster Data

The `ggplot()` function can be used to make maps with raster data as well as vector data. However, `ggplot()` only works with inputs in the form of a data frame and therefore cannot be used to directly map a `SpatRaster` object. Fortunately, raster objects can be converted to data frames with one row per cell, columns for the x and y coordinates, and columns for one or more attributes that are associated with each cell.

In most situations, data frames are very inefficient for storing raster data. Because raster cells are all the same size and arranged on a regular grid, it is not necessary to record the x and y coordinates of every cell. If the coordinates of just one cell (usually called the origin) are known, then the coordinates of any other cell can be determined automatically based on its relative position in the grid. In a data frame, we are effectively storing a vector representation of the raster dataset where each cell is stored as a separate point feature with its own x and y coordinates. Spatial data in a raster object take up less space and can usually be processed more efficiently than the same data stored as points in a data frame. However, converting rasters to data frames will allow us to use `ggplot()` for all our maps and charts and is easier at this stage than learning a different graphing system just for rasters.

Converting raster data to a data frame is straightforward but requires several lines of code. To automate the process, a custom function can be used. The custom function `rasterdf()` takes a required argument, x, which is the raster object to convert to a data frame. The code in the function extracts the x and y coordinates and data values from the raster and then combines them into a data frame that contains the coordinates and the data. It also has an optional argument, `aggregate`, that can be used to downsample the raster resolution before converting. In many situations, the raster dataset may contain more cells than the resolution of the image file or screen that will be used to display the map. Using the `aggregate` argument to downscale the image before converting can greatly reduce the size of the data frame and speed up the mapping process.

The code below uses the `function()` function to create a custom function.

```
rasterdf <- function(x, aggregate = 1) {
  resampleFactor <- aggregate
  inputRaster <- x
  inCols <- ncol(inputRaster)
  inRows <- nrow(inputRaster)
  # Compute numbers of columns and rows in the resampled raster
  resampledRaster <- rast(ncol=(inCols / resampleFactor),
                          nrow=(inRows / resampleFactor),
                          crs = crs(inputRaster))
  # Match to the extent of the original raster
  ext(resampledRaster) <- ext(inputRaster)
  # Resample data on the new raster
  y <- resample(inputRaster,resampledRaster,method='near')
  # Extract cell coordinates into a data frame
  coords <- xyFromCell(y, seq_len(ncell(y)))
```

```
# Extract layer names
dat <- stack(values(y, dataframe = TRUE))
# Add names - 'value' for data, 'variable' for different
# layer names in a multilayer raster
names(dat) <- c('value', 'variable')
dat <- cbind(coords, dat)
dat
}
```

The result is written to a new object called `rasterdf`, which belongs to the `function` class.

```
class(rasterdf)
## [1] "function"
```

The custom `rasterdf()` function can be used to convert a `SpatRaster` object into a data frame. This function can be used in any script, but the code to create the function must always be run before actually calling the function to convert data. The following code converts the LST raster to a data frame after aggregating the cell values by a factor of three.

```
lst_df <- rasterdf(lst_c, aggregate = 3)
summary(lst_df)
##        x                 y                value
##  Min.   :-104.42   Min.   :30.01   Min.   :11.23
##  1st Qu.: -95.64   1st Qu.:32.50   1st Qu.:28.05
##  Median : -86.86   Median :35.00   Median :29.55
##  Mean   : -86.86   Mean   :35.00   Mean   :30.29
##  3rd Qu.: -78.07   3rd Qu.:37.50   3rd Qu.:31.53
##  Max.   : -69.29   Max.   :39.99   Max.   :54.77
##                                    NA's   :217276
##        variable
##  temperature:481370
##
##
##
##
##
##
```

The `ggplot()` function can now be used to map the LST raster. The `geom_raster()` function is used with the x and y arguments assigned to the corresponding columns in `lst_df`. The `fill` argument is the `value` column from `lst_df`, which contains the LST values.

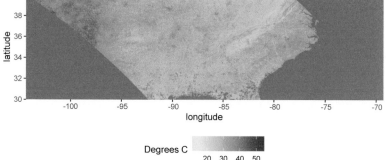

FIGURE 6.1
MODIS Land surface temperature map.

Most of the other `ggplot()` functions have been covered in previous examples.
The `coord_sf()` function is called with the argument `expand = FALSE` to elimi-
nate extra space and the edges of the map. The `legend.position()` argument
to the `theme()` function is used to place the legend at the bottom of the map
(Figure 6.1).

```
ggplot(data = lst_df) +
  geom_raster(aes(x = x,
                  y = y,
                  fill = value)) +
  scale_fill_gradient(name = "Degrees C",
                      low = "yellow",
                      high = "red") +
  coord_sf(expand = FALSE) +
  labs(title = "LST-Adjusted Temperature (Terra Day)",
       x = "longitude",
       y = "latitude") +
  theme(legend.position = "bottom")
```

The **terra** package includes functions for modifying the geometry of raster
datasets. To clip out a smaller portion of the LST dataset, we can use the
`crop()` function. In this example, a bounding box of geographic coordinates
(latitude and longitude) is specified to create a `SpatExtent` object and crop the
raster to the U.S. state of Georgia.

```
clipext <- ext(-86, -80.5, 30, 35.5)
class(clipext)
```

```
## [1] "SpatExtent"
## attr(,"package")
## [1] "terra"
lst_clip <- crop(lst_c, clipext)
lst_clip_df <- rasterdf(lst_clip)
```

There are numerous other **terra** functions for raster GIS operations, including aggregating and resampling to different cell sizes, focal statistics, and zonal statistics. Many of these functions will be demonstrated in the upcoming chapters

Often, it is helpful to plot boundary features on top of a raster to provide context. Vector and raster data can be combined in ggplot() by using the appropriate geom function for each data type. Here, the Georgia county boundaries are overlaid on the LST raster.

```
ga_sf <- st_read(dsn = "GA_SHP.shp", quiet = TRUE)
```

To add the county boundaries, the geom_sf() function can be used as in the examples from the previous chapter. The fill = NA argument displays the polygons as a hollow wireframe so that the raster data can be seen underneath (Figure 6.2).

```
ggplot() +
  geom_raster(data = lst_clip_df,
              aes(x = x,
                  y = y,
                  fill = value)) +
  geom_sf(data=ga_sf,
          color = "grey50",
          fill = NA, size = 0.5) +
  scale_fill_gradient(name = "Degrees C",
                      low = "yellow",
                      high = "red") +
  coord_sf(expand = FALSE) +
  labs(title = "LST-Adjusted Temperature (Terra Day)",
       x = "longitude",
       y = "latitude")
```

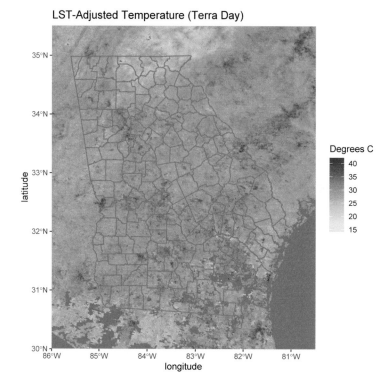

FIGURE 6.2
MODIS Land surface temperature cropped to the state of Georgia with county boundary overlaid.

6.3   Multilayer Rasters

The next examples will use forcings data from the North American Land Data Assimilation System (NLDAS). These are gridded meteorological data that can be obtained through the National Aeronautics and Space Administration's Goddard Earth Sciences Data and Information Services Center (https://disc.gsfc.nasa.gov/). The data have a grid cell size of 0.125 x 0.125 degrees and are stored as GeoTIFF files and can be imported as SpatRaster objects in the same way as the LST dataset.

```
temp2012 <- rast("NLDAS_FORA0125_M.A201207.002.grb.temp.tif")
temp2013 <- rast("NLDAS_FORA0125_M.A201307.002.grb.temp.tif")
temp2014 <- rast("NLDAS_FORA0125_M.A201407.002.grb.temp.tif")
temp2015 <- rast("NLDAS_FORA0125_M.A201507.002.grb.temp.tif")
```

```
temp2016 <- rast("NLDAS_FORA0125_M.A201607.002.grb.temp.tif")
temp2017 <- rast("NLDAS_FORA0125_M.A201707.002.grb.temp.tif")
```

Multiple SpatRaster objects can be combined to produce multi-layer raster datasets. These objects are frequently used for space-time data because they can store rasters from different dates. Multi-layer rasters are created using the c() (combine) function, similar to the way that this function is used in base R to combine multiple objects into vectors. The following code creates two multi-layer SpatRaster objects, one containing July temperature data from 2012–2017 and one containing July precipitation data from 2012–2017. The names() function is used to specify the layer names for each object.

```
tempstack <- c(temp2012, temp2013, temp2014,
               temp2015, temp2016, temp2017)
names(tempstack) <- c("July 2012", "July 2013", "July 2014",
                      "July 2015", "July 2016", "July 2017")
```

Alternatively, multi-layer rasters can be created by making a single call to rast() and providing a vector of file names to import and combine.

```
tempstack <- rast(c("NLDAS_FORA0125_M.A201207.002.grb.temp.tif",
                    "NLDAS_FORA0125_M.A201307.002.grb.temp.tif",
                    "NLDAS_FORA0125_M.A201407.002.grb.temp.tif",
                    "NLDAS_FORA0125_M.A201507.002.grb.temp.tif",
                    "NLDAS_FORA0125_M.A201607.002.grb.temp.tif",
                    "NLDAS_FORA0125_M.A201707.002.grb.temp.tif"))
names(tempstack) <- c("July 2012", "July 2013", "July 2014",
                      "July 2015", "July 2016", "July 2017")
```

When the global() function is applied to multi-layer rasters, each layer in the SpatRaster object is summarized individually.

```
global(tempstack, stat = "mean", na.rm=T)
##              mean
## July 2012 297.8852
## July 2013 296.1879
## July 2014 295.7905
## July 2015 296.2655
## July 2016 296.5719
## July 2017 296.7618
```

From these summaries, is apparent that the temperature units are Kelvin. As was done with the MODIS LST data, these data can be converted to Celsius

using a simple R expression. The expression is automatically applied to all the
layers in the multi-layer `SpatRaster` object.

```
tempstack <- tempstack - 273.15
global(tempstack, stat = "mean", na.rm=T)
##                       mean
## July 2012 24.73522
## July 2013 23.03793
## July 2014 22.64051
## July 2015 23.11549
## July 2016 23.42190
## July 2017 23.61176
```

Multi-layer rasters can be mapped using `ggplot()`. First, the `SpatRaster` object
is converted to a data frame using the custom `rasterdf()` function, and a
dataset of U.S. state boundaries is imported for use as an overlay.

```
tempstack_df <- rasterdf(tempstack)
summary(tempstack_df)
##        x                 y                 value
## Min.   :-124.94   Min.   :25.06    Min.   : 3.09
## 1st Qu.:-110.47   1st Qu.:32.03    1st Qu.:19.59
## Median : -96.00   Median :39.00    Median :23.53
## Mean   : -96.00   Mean   :39.00    Mean   :23.43
## 3rd Qu.: -81.53   3rd Qu.:45.97    3rd Qu.:27.62
## Max.   : -67.06   Max.   :52.94    Max.   :40.18
##                                    NA's   :140982
##          variable
## July 2012:103936
## July 2013:103936
## July 2014:103936
## July 2015:103936
## July 2016:103936
## July 2017:103936
##
states_sf <- read_sf("conterminous_us_states.shp", quiet = TRUE)
```

Then, `ggplot()` is used with `facet_wrap()` to display a composite plot with
one year mapped in each facet. The `variable` column, which contains the
names of the different raster layers, is specified in the `facets` argument. The
temperature data are displayed using a yellow to red color ramp with a vector
dataset of state boundaries overlaid to provide a spatial reference (Figure 6.3).

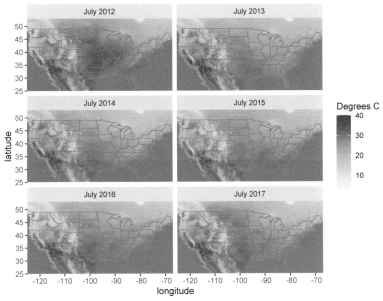

FIGURE 6.3
July temperatures from the NLDAS forcings dataset.

```
ggplot() +
  geom_raster(data = tempstack_df,
              aes(x = x,
                  y = y,
                  fill = value)) +
  geom_sf(data = states_sf,
          fill = NA,
          color = "grey50",
          size = 0.25) +
  scale_fill_gradient(name = "Degrees C",
                      low = "yellow",
                      high = "red") +
  coord_sf(expand = FALSE) +
  facet_wrap(facets = vars(variable), ncol = 2) +
  labs(title = "Mean July Temperature",
       x = "longitude",
       y = "latitude")
```

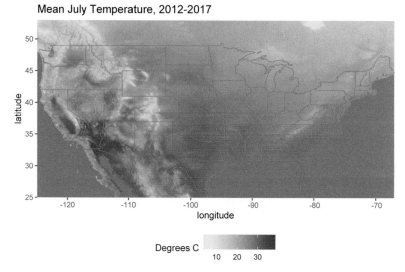

FIGURE 6.4
Mean July temperatures calculated from 2012–2017.

6.4 Computations on Raster Objects

Many statistical summary functions have methods for raster objects. When used with multi-layer rasters, these functions will be applied to each pixel to summarize the data across all layers. For example, the following code generates a mean temperature for each pixel using the six years of data stored in different layers. Note the difference compared to the summaries calculated with `global()`, which were summarized across all pixels for each layer instead of across all layers for each pixel.

```
meantemp <- mean(tempstack)
```

As in the previous example, the temperature is displayed using a yellow to red color ramp with a vector dataset of state boundaries overlaid (Figure 6.4).

```
meantemp_df <- rasterdf(meantemp)

ggplot() +
  geom_raster(data = meantemp_df, aes(x = x,
                                      y = y,
                                      fill = value)) +
```

```
geom_sf(data = states_sf,
        fill = NA,
        color = "grey50",
        size = 0.25) +
scale_fill_gradient(name = "Degrees C",
                    low = "yellow",
                    high = "red") +
coord_sf(expand = FALSE) +
labs(title = "Mean July Temperature, 2012-2017",
     x = "longitude", y = "latitude") +
theme(legend.position = "bottom")
```

The following code subtracts each of the six annual datasets in the multi-layer raster from the mean temperature layer to generate maps of temperature anomalies. In this example, `tempstack` is a raster object with six layers, and `meantemp` is a raster object with one layer. When the subtraction operator is used, `meantemp` is subtracted from each layer of `tempstack`, producing a new `SpatRaster` object with six layers called `tempanom`. The layer names from `tempstack` are then assigned to `tempanom`.

```
tempanom <- tempstack - meantemp
names(tempanom) <- names(tempstack)
```

The anomalies are displayed using a bicolor ramp generated with the `scale_fill_gradients2()` function. This approach is justified because the anomalies have a breakpoint at zero. Positive (warm) anomalies are displayed as red, and negative (cold) anomalies are displayed as blue. By default, the color break is at zero, but this value can be changed by modifying the `'midpoint` argument to `scale_fill_gradient2()`. The `theme_void()` function is used to remove the axes and graticules from the maps, and the `theme()` function is used to bold the text labels and increase the font sizes. Using the `expand = TRUE` argument with `coord_df()` provides some additional space around the maps and makes their labels easier to read (Figure 6.5).

```
tempanom_df <- rasterdf(tempanom)

ggplot() +
  geom_raster(data = tempanom_df, aes(x = x,
                                      y = y,
                                      fill = value)) +
  geom_sf(data = states_sf,
          fill = NA,
          color = "grey50",
```

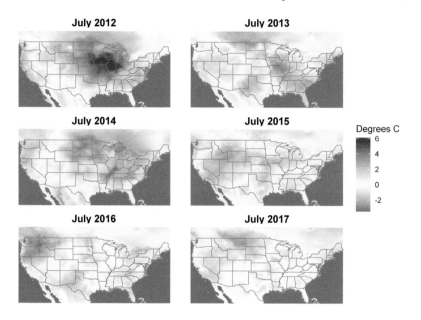

FIGURE 6.5
July temperature anomalies from the NLDAS forcings dataset.

```
        size = 0.25) +
  scale_fill_gradient2(name = "Degrees C",
                       low = "blue",
                       mid = "lightyellow",
                       high = "red") +
  coord_sf(expand = TRUE) +
  facet_wrap(facets = vars(variable), ncol = 2) +
  theme_void() +
  theme(strip.text.x = element_text(size=12, face="bold"))
```

In most real applications, more than six years of meteorological data (typically 30 years) would be used to generate a mean climatology for calculating anomalies. This small example is presented to illustrate the advantages of converting annual meteorological data into anomalies. The untransformed temperature data show the same strong spatial patterns every year. By subtracting the mean temperature map from each annual temperature map, the spatial pattern of temperature is removed and the interannual variation is emphasized in the anomaly maps.

One or more layers can be extracted from a raster object using double brackets. As with lists in base R, raster layers can be extracted either by number or name.

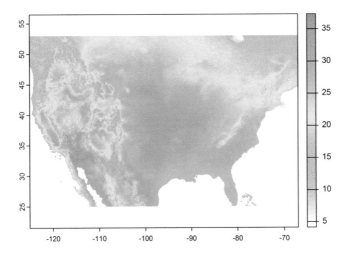

FIGURE 6.6
Raster map created with the generic plot method.

```
temp1 <- tempstack[[1]]
names(temp1)
## [1] "July 2012"
temp2 <- tempstack[["July 2012"]]
names(temp2)
## [1] "July 2012"
temp3 <- tempstack[[1:3]]
names(temp3)
## [1] "July 2012" "July 2013" "July 2014"
```

The **terra** package has *methods* for many of the base R plotting functions. This means that a SpatRaster object can be provided as an input to these functions, and R will be able to generate an appropriate map or graph. For example, the generic plot() function is handy for generating a quick map of a raster object (Figure 6.6).

```
plot(tempstack[[1]])
```

The plot() function will also work with multi-layer rasters, automatically generating a multi-panel figure with one map for each layer (Figure 6.7).

```
plot(tempstack)
```

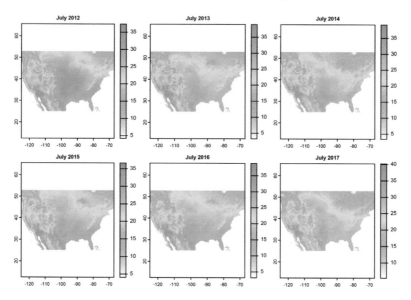

FIGURE 6.7
Multilayer raster map created with the generic plot method.

The `hist()` function will automatically generate a histogram or density plot for each layer in a raster object (Figure 6.8).

```
hist(tempstack)
```

When two single-layer rasters are provided as arguments to the `plot()` function, it will generate a scatterplot. Because of the large number of cells in most raster datasets, simply plotting them all as points produces a large blob. Setting the `gridded=TRUE` argument produces a gridded scatterplot, in which the value of each grid cell represents the density of points in that portion of the scatterplot (Figure 6.9).

```
plot(temp2012, temp2013,
     xlab = "2012",
     ylab = "2013",
     gridded = TRUE)
```

These functions are very handy for generating quick, interactive visualizations for large raster objects, and it is possible to customize them more if you are familiar with plotting in base R. However, it is also possible to convert raster data into a data frame using `rasterdf()` and graph it with `ggplot()` if you want to have more control over the layout of your plot. Several examples

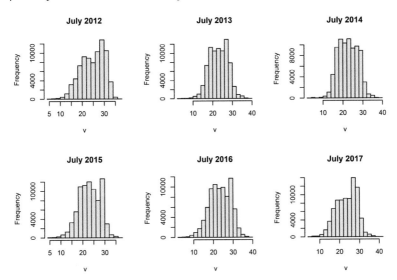

FIGURE 6.8
Histograms of raster layers.

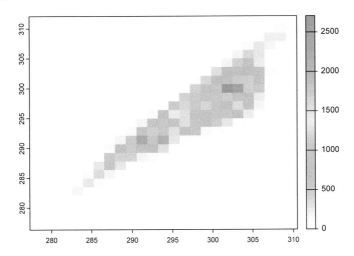

FIGURE 6.9
Gridded scatterplot of two raster datasets.

of using `ggplot()` to plot summaries of raster data will be provided in later chapters. There are also a number of specialized mapping packages that work with **terra** and **sf** objects, including **rasterVis** (Perpinan Lamigueiro and Hijmans, 2022), **tmap** (Tennekes, 2018), and **leaflet** (Cheng et al., 2022).

More resources on the *terra* package can be found at `https://rspatial.org/t erra/spatial/index.html`, including detailed explanations of many functions and examples with a variety of remote sensing datasets. The upcoming second edition of *Geocomputation with R* by Robin Lovelace and others (Lovelace et al., 2019) will incorporate the **terra** packages and should be an excellent reference and source of further examples.

6.5   Practice

1. Convert the clipped LST raster from Celsius to Fahrenheit and generate an updated version of the Georgia LST map.

2. Generate a new raster where each pixel contains the standard deviation of the NLDAS July Temperature values from 2012–2017. Map the result to explore where temperatures were most variable over this period.

3. Create a new raster dataset where each cell contains the difference between July 2015 temperature and July 2016 temperature. Examine the map and histogram of this raster and use them to assess changes in temperature between 2015 and 2016.

7

Raster Geospatial Data—Discrete

The previous chapter covered raster datasets with continuous variables, which potentially have an infinite number of values. This chapter introduces raster datasets with discrete variables, which are classified into a limited number of values. We still use numbers to work with discrete data. For example, different types of land cover can be identified using integers. Water might be coded as 11, built-up areas as 23, and deciduous forests as 41. In this situation, the numbers themselves are not meaningful, so summary statistics such as sum() and mean() would not provide any useful information. Therefore, different approaches are needed when visualizing and summarizing discrete raster data.

This chapter will use classified land cover data from the National Land Cover Database (NLCD). The NLCD produces a standardized land cover data product for the United States. Data are available from 2001–2019 at 2–3 year intervals. These products have been generated from Landsat satellite imagery using consistent methods that make the resulting products suitable for change analysis. More information on the NLCD can be found at the Multi-Resolution Land Characteristics Consortium (MRLC) website http://www.mrlc.gov/. Data can be downloaded as very large files covering the conterminous U.S., Alaska, or U.S. Islands and Territories from the MRLC website. Smaller areas can be selected and downloaded using the MRLC viewer at https://www.mrlc.gov/viewer/. For this exercise, we will work with a small portion of the NLCD encompassing Walton County, GA. This rural county is located on the outskirts of the Atlanta metropolitan area and is under considerable development pressure.

Most of the required packages have been used in previous chapters. The **colorspace** package is new and contains functions for manipulating colors and palettes.

```
library(RColorBrewer)
library(ggplot2)
library(colorspace)
library(dplyr)
library(tidyr)
library(readr)
```

```
library(sf)
library(terra)
```

7.1   Importing and Mapping Land Cover Data

The Walton County data are read from a TIFF file to create a `SpatRaster`
object. The dataset is a grid of 30 m square cells with 1579 rows and 1809
columns.

```
nlcd19 <- rast("NLCD_2019_Land_Cover_Walton.tiff")
class(nlcd19)
## [1] "SpatRaster"
## attr(,"package")
## [1] "terra"
nrow(nlcd19)
## [1] 1579
ncol(nlcd19)
## [1] 1809
res(nlcd19)
## [1] 30 30
```

Mapping these raster data with `ggplot()` requires the `rasterdf()` custom
function to convert them to a data frame. To avoid having to paste the code
for this function into every script, it can be stored in a separate file and run
using the `source()` function. The file containing the function code needs to be
in the working directory, or else the file path needs to be explicitly specified.

```
source("rasterdf.R")
nlcd19_df <- rasterdf(nlcd19)
```

By default, the `geom_raster()` function generates a raster map with a continuous color ramp in shades of blue (Figure 7.1).

```
ggplot(data = nlcd19_df) +
  geom_raster(aes(x = x,
                  y = y,
                  fill = value))
```

This map and the accompanying legend are not particularly useful. The values
stored in the land cover raster are numeric codes. For example, a value of

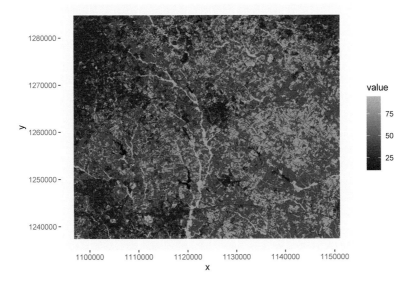

FIGURE 7.1
Land cover map with default color ramp.

11 represents water, whereas a value of 21 represents low-density residential development. The default chart created by `ggplot()` maps these values on a continuous scale rather than as categories, and the resulting symbology and legend have no real meaning. A detailed list of all the NLCD codes can be found at `https://www.mrlc.gov/data/legends/national-land-cover-database-class-legend-and-description`.

To generate a better-looking map, it will be necessary to assign each land cover code in the map a unique name and color for display. The `unique()` function can be used to extract all of the numerical land cover codes that are present in the Walton County dataset. The function output is a data frame with one column, which is extracted using subscripts.

```
LCcodes <- unique(nlcd19)[, 1]
LCcodes
## [1] 11 21 22 23 24 31 41 42 43 52 71 81 82 90 95
```

Although the NLCD contains sixteen different land cover types for the conterminous United States, there are only fifteen in Walton County. The ice and snow class (12) is missing. Another vector is created with the corresponding human-readable names for each land cover class.

```
LCnames <-c(
  "Water",
  "DevelopedOpen",
  "DevelopedLow",
  "DevelopedMed",
  "DevelopedHigh",
  "Barren",
  "DeciduousForest",
  "EvergreenForest",
  "MixedForest",
  "ShrubScrub",
  "GrassHerbaceous",
  "PastureHay",
  "CultCrops",
  "WoodyWetlands",
  "EmergentHerbWet")
```

Next, the `coltab()` function is used to extract color table information from
the `SpatRaster` object. This color table is embedded in the NLCD TIFF data
files and was read in and stored along with the data. Not all raster datasets
include a color table, but we can take advantage of it here to display the land
cover classes using the "official" NLCD color scheme. Color information in
the red, green, and blue channels is extracted with the `coltab()` function and
converted to a data frame. Then, the rows associated with the NLCD land
cover codes are extracted by subscripting. Finally, the `rgb()` function is used
to convert the red, green, and blue information into hexadecimal color codes.

```
nlcdcols <- data.frame(coltab(nlcd19))
nlcdcols <- nlcdcols[LCcodes + 1,]
LCcolors <- rgb(red = nlcdcols$red,
                green = nlcdcols$green,
                blue = nlcdcols$blue,
                names = as.character(LCcodes),
                maxColorValue = 255)
LCcolors
##          11        21        22        23        24        31
## "#466B9F" "#DEC5C5" "#D99282" "#EB0000" "#AB0000" "#B3AC9F"
##          41        42        43        52        71        81
## "#68AB5F" "#1C5F2C" "#B5C58F" "#CCB879" "#DFDFC2" "#DCD939"
##          82        90        95
## "#AB6C28" "#B8D9EB" "#6C9FB8"
```

Now the `ggplot()` function can be used to plot the NLCD land cover raster. In
the `aes()` function, the raster value is converted to a character so that `ggplot()`

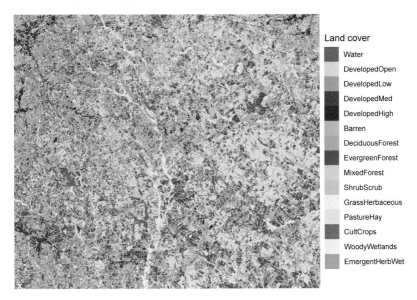

FIGURE 7.2
Land cover map with unique colors and labels for every land cover class.

will recognize it as a categorical variable. The `scale_fill_manual()` function is used to specify the colors that match the NLCD codes. The `na.translate` argument is specified as `FALSE` so that an `NA` value does not show up in the legend (Figure 7.2).

```
ggplot(data = nlcd19_df) +
  geom_raster(aes(x = x,
                  y = y,
                  fill = as.character(value))) +
  scale_fill_manual(name = "Land cover",
                    values = LCcolors,
                    labels = LCnames,
                    na.translate = FALSE) +
  coord_sf(expand = FALSE) +
  theme_void()
```

In the last chapter, the size of a raster dataset was reduced by defining a new extent and then using the `crop()` function. It is also possible to zoom in and map a portion of a raster dataset without modifying the underlying data. Here, a subset of Walton County is mapped by setting the x and y limits of the plot using the `coord_sf()` function (Figure 7.3).

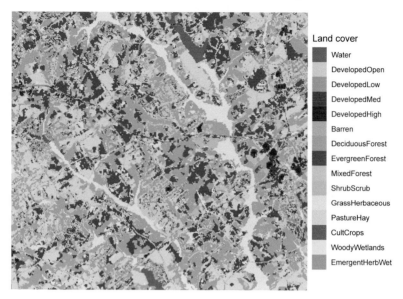

FIGURE 7.3
Zoomed-in land cover map with unique colors and label for every land cover
class.

```
ggplot(data = nlcd19_df) +
  geom_raster(aes(x = x,
                  y = y,
                  fill = as.character(value))) +
  scale_fill_manual(name = "Land cover",
                    values = LCcolors,
                    labels = LCnames,
                    na.translate = FALSE) +
  coord_sf(expand = FALSE,
           xlim = c(1114000, 1125000),
           ylim = c(1260000, 1270000)) +
  theme_void()
```

7.2 Reclassifying Raster Data

Often, land cover patterns can be seen more clearly if the land cover classes
are aggregated into a smaller number of broader land cover types. After listing
the classes in the Walton County dataset, a vector is created with the codes

for the new land cover classes. These two vectors are combined to create a
lookup table that associates the old and new codes. The classify() function
is then used to assign a new land cover class to every pixel in Walton County.
Vectors are also created with the names and display colors for the new classes.

```
LCcodes
##  [1] 11 21 22 23 24 31 41 42 43 52 71 81 82 90 95
newclas <- c(1, 2, 2, 2, 2, 3, 4, 4, 4, 5, 5, 5, 6, 7, 7)
lookup <- data.frame(LCcodes, newclas)
nlcd19_rc <- classify(nlcd19, lookup)
newnames <- c("Water",
              "Developed",
              "Barren",
              "Forest",
              "GrassShrub",
              "Cropland",
              "Wetland")
newcols <- c("mediumblue",
             "firebrick2",
             "gray60",
             "darkgreen",
             "yellow2",
             "orange4",
             "paleturquoise2")
```

The same ggplot() code as before can be used to generate a categorical map with the new reclassified raster dataset (Figure 7.4). In the
scale_fill_manual() function, the new colors and class names for the reclassified data are specified for the values and labels arguments.

```
nlcd19_rc_df <- rasterdf(nlcd19_rc)

ggplot(data = nlcd19_rc_df) +
  geom_raster(aes(x = x, y = y, fill = as.character(value))) +
  scale_fill_manual(name = "Land cover",
                    values = newcols,
                    labels = newnames,
                    na.translate = FALSE) +
  coord_sf(expand = FALSE) +
  theme_void()
```

The colors displayed in the map are heavily saturated—they appear bold and
vivid. However, mixing many bright, saturated colors in a map can make
it difficult to discern the underlying patterns. Thus, it is often useful to

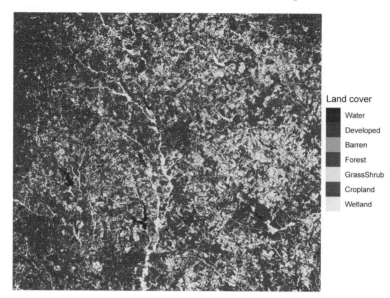

FIGURE 7.4
Land cover map with reclassified land cover types.

desaturate the colors used in maps. Desaturating mixes gray into the colors to make them appear softer and more muted. A simple way to do this is to use the desaturate() function from the **colorspace** package. The amount argument controls how much saturation is reduced for all the colors in the palette.

```
newcols2 <- desaturate(newcols, amount = 0.4)
```

The map of the reclassified land cover data can now be recreated using the desaturated color palette (Figure 7.5).

```
ggplot(data = nlcd19_rc_df) +
  geom_raster(aes(x = x,
                  y = y,
                  fill = as.character(value))) +
  scale_fill_manual(name = "Land cover",
                    values = newcols2,
                    labels = newnames,
                    na.translate = FALSE) +
  coord_sf(expand = FALSE) +
  theme_void()
```

Another useful visualization technique is the "small multiples" approach, in which each land cover class is displayed in a separate map. To generate this

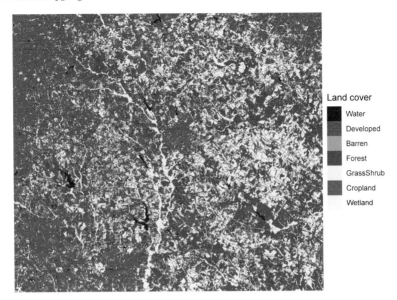

FIGURE 7.5
Land cover map with reclassified land cover types and desaturated colors.

map, a separate binary raster must be generated for each of the land cover
types. One way to do this is with a logical statement. The following example
converts the reclassified land cover raster into a new raster where cells belonging
to the developed class have a value of TRUE (1) and all other cells have a value
of FALSE (0).

```
developed <- nlcd19_rc == 2
summary(developed)
##      Layer_1
## Min.    :0.000
## 1st Qu.:0.000
## Median :0.000
## Mean    :0.244
## 3rd Qu.:0.000
## Max.    :1.000
```

To convert multiple classes, the segregate() function can be used to convert
every unique class in a discrete raster to a 1/0 binary raster. Because there are
seven reclassified land cover types, the output of segregate() is a SpatRaster
object with seven layers ordered by the land cover codes. Subscripting is used
to remove the water layer from the SpatRaster object and the vector of class
names.

```
nlcd19_stk <- segregate(nlcd19_rc)
nlcd19_stk <- nlcd19_stk[[-1]]
names(nlcd19_stk) <- newnames[-1]
```

As demonstrated in the previous chapter, the `rasterdf()` function can be used to convert multi-layer rasters as well as single-layer rasters into data frames for mapping with `ggplot()`. In the resulting data frame, the `variable` column contains the name of each raster layer in the stack.

```
nlcd19_stk_df <- rasterdf(nlcd19_stk)
summary(nlcd19_stk_df)
##       x                y                value
## Min.   :1096830   Min.   :1237410   Min.   :0.0000
## 1st Qu.:1110390   1st Qu.:1249230   1st Qu.:0.0000
## Median :1123950   Median :1261080   Median :0.0000
## Mean   :1123950   Mean   :1261080   Mean   :0.1645
## 3rd Qu.:1137510   3rd Qu.:1272930   3rd Qu.:0.0000
## Max.   :1151070   Max.   :1284750   Max.   :1.0000
##         variable
## Developed :2856411
## Barren    :2856411
## Forest    :2856411
## GrassShrub:2856411
## Cropland  :2856411
## Wetland   :2856411
```

These data can be used to generate a series of binary maps - one for each of the six land cover classes (Figure 7.6). Mapping each class separately makes it easier to see the rare classes, such as barren land and cropland. Patterns in some of the other classes, such as the higher concentration of developed land in the eastern part of the map, are also easier to discern than in the multiclass map.

```
ggplot(data = nlcd19_stk_df) +
  geom_raster(aes(x = x,
                  y = y,
                  fill = as.character(value))) +
  scale_fill_manual(name = "Present",
                    values = c("gray80", "gray20"),
                    labels = c("No", "Yes"),
                    na.translate = FALSE) +
  facet_wrap(facets = vars(variable), ncol = 3) +
  coord_sf(expand = FALSE) +
```

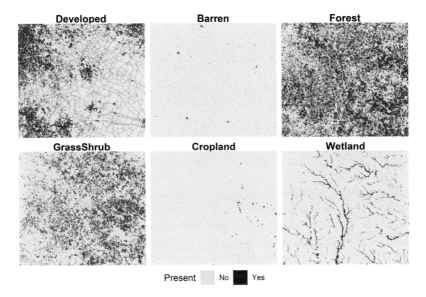

FIGURE 7.6
Land cover map with reclassified land cover types displayed as small multiples.

```
theme_void() +
theme(strip.text.x = element_text(size=12, face="bold"),
      legend.position="bottom")
```

7.3 Focal Analysis of Raster Data

Focal analysis computes a summary of all the cells within a window surrounding an individual cell and assigns the value of the summary to that cell. This operation is repeated for all cells in a raster dataset, producing a new raster where every cell contains a focal summary. In most cases, the result of a focal analysis is a smoothed version of the input raster, where variation at smaller scales than the summary window is removed and broader scale spatial patterns are emphasized.

To carry out a focal analysis, a weights object must be defined using the `focalWeight()` function. Here, a range of circular windows with radii from 100 m to 2000 m are used. By default, the weight of each cell within this window is equal to its proportion of the total window.

```
forest <- nlcd19_stk[["Forest"]]
fwts100 <- focalMat(forest, d=100, type = "circle")
fwts500 <- focalMat(forest, d=500, type = "circle")
fwts1000 <- focalMat(forest, d=1000, type = "circle")
fwts2000 <- focalMat(forest, d=2000, type = "circle")
```

Then the `focal()` function is called with the weights objects and a summary
function as arguments. By summing across the weights, the `focal()` function
computes the proportion of each land cover class within the specified windows.

```
for_100 <- focal(forest, w=fwts100, fun=sum)
for_500 <- focal(forest, w=fwts500, fun=sum)
for_1000 <- focal(forest, w=fwts1000, fun=sum)
for_2000 <- focal(forest, w=fwts2000, fun=sum)
focal_stk <- c(for_100, for_500, for_1000, for_2000)
names(focal_stk) <- c("100m", "500m", "1000m", "2000m")
```

Changing the size of the focal window changes the degree of smoothing in
the resulting maps, similar to the bandwidth of a kernel density analysis
(Figure 7.7).

```
focal_stk_df <- rasterdf(focal_stk)

ggplot(data = focal_stk_df) +
  geom_raster(aes(x = x,
                  y = y,
                  fill = value)) +
  scale_fill_distiller(name = "Density",
                       palette = "YlOrRd") +
  facet_wrap(facets = vars(variable), ncol = 2) +
  coord_sf(expand = TRUE) +
  theme_void() +
  theme(strip.text.x = element_text(size=12, face="bold"),
        legend.position="bottom")
```

By default, the `focal()` function will return a value of NA if there are any NA
values inside the window or if the window extends outside the boundary of the
raster dataset. These NA values are displayed in gray in the preceding figure,
and the gray boundary around the outside of the dataset increases in width as
the window size increases.

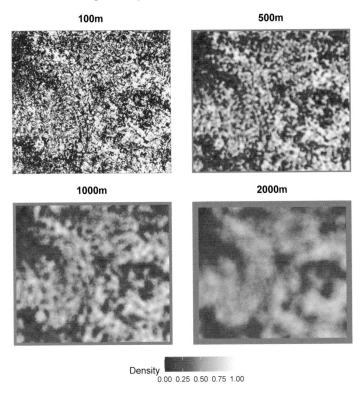

FIGURE 7.7
Forest density summarized as a focal mean within different-sized circular windows.

7.4 Land Cover Change Analysis

Land cover patterns change over time as a result of urban growth, agricultural abandonment, clearcutting, and other land use activities. These trends can be quantified and analyzed using multiple years of NLCD data. The following code imports eight NLCD data files, spanning 2001–2019, into a multilayer raster dataset. Then the data are reclassified into the seven broader land cover classes using the `classify()` function.

```
nlcd_stk <- rast(c("NLCD_2001_Land_Cover_Walton.tiff",
                   "NLCD_2004_Land_Cover_Walton.tiff",
                   "NLCD_2006_Land_Cover_Walton.tiff",
                   "NLCD_2008_Land_Cover_Walton.tiff",
                   "NLCD_2011_Land_Cover_Walton.tiff",
```

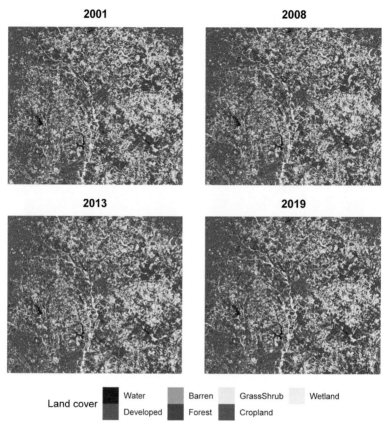

FIGURE 7.8
Land cover map with four years of data.

```
                    "NLCD_2013_Land_Cover_Walton.tiff",
                    "NLCD_2016_Land_Cover_Walton.tiff",
                    "NLCD_2019_Land_Cover_Walton.tiff"))

nlcd_rc <- classify(nlcd_stk, lookup)
names(nlcd_rc) <- c("2001", "2004", "2006", "2008", "2011",
                    "2013", "2016", "2019")
```

The ggplot() function can be used with facet_wrap() to map the time series of
land cover (Figure 7.8). Before it is converted to a data frame with rasterdf(),
the multilayer raster object is subscripted with double brackets to select four
years for visualization.

```
nlcd_rc_df  <- rasterdf(nlcd_rc[[c("2001", "2008",
                                   "2013", "2019")]])


ggplot(data = nlcd_rc_df) +
  geom_raster(aes(x = x,
                  y = y,
                  fill = as.character(value))) +
  scale_fill_manual(name = "Land cover",
                    values = newcols2,
                    labels = newnames,
                    na.translate = FALSE) +
  coord_sf(expand = TRUE) +
  facet_wrap(facets = vars(variable), ncol = 2) +
  theme_void() +
  theme(strip.text.x = element_text(size = 12, face="bold"),
        legend.position="bottom")
```

When working with land cover data, it is usually important to know which land cover classes are increasing or decreasing and by how much. By looking closely at these maps, it is possible to see areas where land cover is changing. However, it is not easy to discern the overall trends. One way to visualize the changes more effectively is to calculate the total area of each land cover class in each year and plot the changes over time.

The rectangular boundary of the raster dataset includes Walton County as well as other surrounding areas that fit within the bounding box. The Walton County boundaries are extracted from the same shapefile of Georgia counties that was introduced in Chapter 6. These data are reprojected to be in the same coordinate reference system as the NLCD data.

```
gacounty <- st_read("GA_SHP.shp", quiet = TRUE)
gacounty <- st_transform(gacounty, crs(nlcd_rc))
walton <- filter(gacounty, NAME10 == "Walton")
```

To restrict the analysis to only Walton county, the crop() and mask() functions are used to trim the size of the raster to the borders of Walton County and convert all pixels outside of Walton County to NA values. When vector data are supplied as an argument to a **terra** function, they must be converted to objects of the class SpatVector(). This conversion is accomplished using the vect() function.

```
nlcd_rc_crp <- crop(nlcd_rc, vect(walton))
nlcd_rc_msk <- mask(nlcd_rc_crp, vect(walton))
```

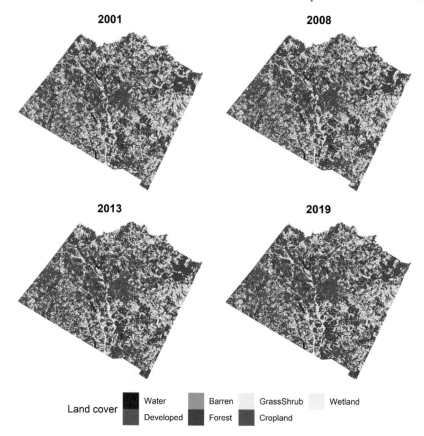

FIGURE 7.9
Land cover map with four years of data cropped and masked to the Walton
County boundaries.

The cropped and masked dataset now only contains land cover data within
the boundaries of Walton County (Figure 7.9).

```
nlcd_msk_df <- rasterdf(nlcd_rc_msk[[c("2001", "2008",
                                       "2013", "2019")]])


ggplot(data = nlcd_msk_df) +
  geom_raster(aes(x = x,
                  y = y,
                  fill = as.character(value))) +
  scale_fill_manual(name = "Land cover",
                    values = newcols2,
                    labels = newnames,
```

```
                      na.translate = FALSE) +
  coord_sf(expand = TRUE) +
  facet_wrap(facets = vars(variable), ncol = 2) +
  theme_void() +
  theme(strip.text.x = element_text(size=12, face="bold"),
        legend.position="bottom")
```

The freq() function can be used to extract the number of cells in each land
cover class from each layer in the raster stack. The output is a data frame
with one row for each combination of year and land cover class.

```
freq_df <- freq(nlcd_rc_msk, usenames=TRUE)
glimpse(freq_df)
## Rows: 56
## Columns: 3
## $ layer <chr> "2001", "2001", "2001", "2001", "2001", "200~
## $ value <dbl> 1, 2, 3, 4, 5, 6, 7, 1, 2, 3, 4, 5, 6, 7, 1,~
## $ count <dbl> 9669, 127218, 3674, 458074, 299985, 2707, 51~
```

Before plotting the data frame, a series of **dplyr** functions is used to modify the
columns. The count of 30-meter square cells is converted to square kilometers.
The value column is converted to a new factor column called class with labels
corresponding to the class names. Using the class column will ensure that the
land cover class will be treated like a categorical variable instead of a numeric
variable and that the appropriate label for each class will be displayed in the
graphs. The layer column, which contains the name of each raster layer, is
converted to a numeric column containing the year value.

```
nlcd_chg <- freq_df %>%
  mutate(km2 = count * 900 / 1000000,
         class = factor(value,
                        levels = 1:7,
                        labels = newnames),
         year = as.numeric(layer))
```

The change in area over time for each class can be displayed as a line graph
(Figure 7.10). The lines for different land cover map classes are distinguished by
color, and the same colors that were used in the land cover maps are assigned
for consistency. As shown in Chapter 5, the expression() function is used to
generate a superscript. To include the final parenthesis, a star (*) is required
after the superscript expression to indicate that it will be followed by more
text.

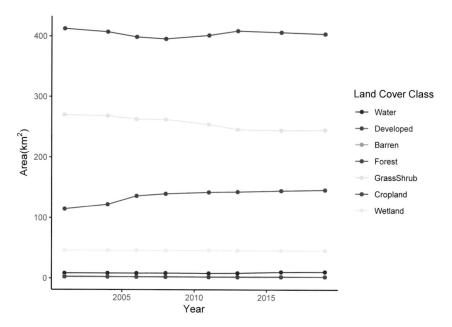

FIGURE 7.10
Change in area of land cover classes in Walton County displayed as a line graph.

```
ggplot(data = nlcd_chg) +
  geom_line(aes(x = year, y = km2, color = class)) +
  geom_point(aes(x = year, y = km2, color = class)) +
  scale_color_manual(name = "Land Cover Class",
                     values = newcols) +
  labs(x = "Year", y = expression("Area(km"^2*")")) +
  theme_classic()
```

Another way to graph the changes is using `facet_wrap()` to create a separate subplot for each land cover class (Figure 7.11).

```
ggplot(data = nlcd_chg) +
  geom_line(aes(x = year, y = km2)) +
  facet_wrap(facets = vars(class), ncol = 4) +
  labs(x = "Year", y = expression("Area(km"^2*")")) +
  theme_bw()
```

In both of the previous graphs, it is very difficult to see the trends of the less common land cover classes. To see their change more clearly, the scale of the

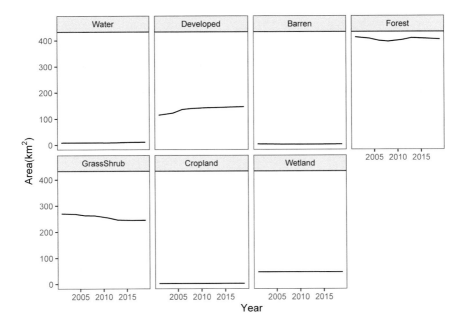

FIGURE 7.11
Change in area of land cover classes in Walton County displayed as a faceted
line graph.

y-axis can be changed to vary freely with the range of values for each class
(Figure 7.12).

```
ggplot(data = nlcd_chg) +
  geom_line(aes(x = year, y = km2)) +
  facet_wrap(facets = vars(class),
             scales = "free_y",
             ncol = 4) +
  labs(x = "Year", y = expression("Area(km"^2*")")) +
  theme_bw()
```

7.5   Land Cover Transition Matrices

Another way to quantify land cover change between two years is through
a transition matrix. Generating the transition matrix requires a multilayer
raster containing land cover layers for the beginning (2001) and end (2019)

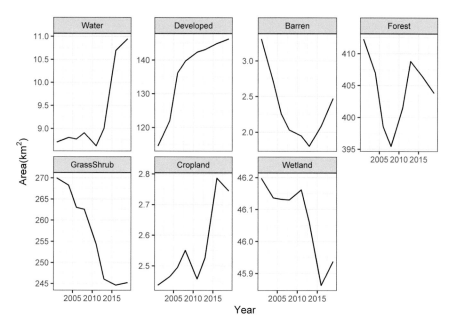

FIGURE 7.12
Change in area of land cover classes in Walton County displayed as a faceted line graph with free scales on the y-axis.

of the change period. The `crosstab()` function is then used to calculate the transition values. This function outputs the results as a `table` object, which is then converted to a tibble (data frame) object.

```
changeras <- c(nlcd_rc_msk[["2001"]],
               nlcd_rc_msk[["2019"]])
changeout <- crosstab(changeras)
class(changeout)
## [1] "xtabs" "table"
changedf <- as_tibble(changeout)
class(changedf)
## [1] "tbl_df"      "tbl"        "data.frame"
```

The data frame has one row for each combination of 2001 and 2019 land cover classes. The columns with the 2001 and 2019 class codes have an `X` added to their names because column names cannot begin with a number. The `Freq` column contains the pixel counts for each combination of 2001 and 2019 land cover classes.

```
changedf
## # A tibble: 49 x 3
##    X2001 X2019      n
##    <chr> <chr>  <int>
## 1 1     1       9136
## 2 2     1          0
## 3 3     1         49
## 4 4     1       2055
## 5 5     1        350
## 6 6     1          0
## 7 7     1        575
## 8 1     2         58
## 9 2     2     127218
## 10 3    2       1428
## # ... with 39 more rows
```

To make the output easier to work with, the data frame is modified using
dplyr functions. The class codes are converted to factors and labeled with
abbreviated class names. A combination of `group_by()` and `mutate()` (instead
of `summarize()`) is used to compute the total area of each class in 2001. When
`sum()` and other summary functions are used inside of the `mutate()` function
after the data have been grouped, they return one value for each row in the
data frame instead of one value per group. Then, the area of each transition is
computed as a percentage of the 2001 area of each land cover class.

```
shortnames <- c("Wat",
                "Dev",
                "Bare",
                "For",
                "Grass",
                "Crop",
                "Wet")

changedf <- changedf %>%
  mutate(X2001 = factor(X2001,
                        levels = 1:7,
                        labels = shortnames),
         X2019 = factor(X2019,
                        levels = 1:7,
                        labels = shortnames),
         ha = n * 900/10000) %>%
  group_by(X2001) %>%
  mutate(tot2001 = sum(ha),
```

TABLE 7.1

Change in hectares from 2001–2019 in Walton County, with 2001 classes as rows and 2019 classes as columns.

	Wat	**Dev**	**Bare**	**For**	**Grass**	**Crop**	**Wet**
Wat	822.2	5.2	0.4	15.3	6.5	0.0	20.6
Dev	0.0	11449.6	0.0	0.0	0.0	0.0	0.0
Bare	4.4	128.5	159.7	28.0	9.8	0.0	0.3
For	184.9	1606.2	45.7	36915.0	2463.1	0.6	11.0
Grass	31.5	1410.6	40.9	3419.9	22040.6	48.7	6.5
Crop	0.0	18.4	0.0	0.0	0.0	225.2	0.1
Wet	51.8	6.9	0.7	1.4	3.7	0.0	4555.4

```
        perc = 100 * ha / tot2001)

changedf
## # A tibble: 49 x 6
## # Groups:   X2001 [7]
##     X2001 X2019      n      ha tot2001    perc
##     <fct> <fct> <int>   <dbl>   <dbl>   <dbl>
##  1 Wat   Wat    9136   822.     870.   94.5
##  2 Dev   Wat       0     0    11450.    0
##  3 Bare  Wat      49    4.41    331.    1.33
##  4 For   Wat    2055   185.   41227.    0.449
##  5 Grass Wat     350    31.5  26999.    0.117
##  6 Crop  Wat       0     0      244.    0
##  7 Wet   Wat     575    51.8   4620.    1.12
##  8 Wat   Dev      58     5.22   870.    0.600
##  9 Dev   Dev  127218 11450.   11450.  100
## 10 Bare  Dev    1428   129.     331.   38.9
## # ... with 39 more rows
```

These transitions are often displayed as a change matrix. This is an n x n matrix, where n is the number of land cover classes (Table 7.1).

```
changemat <- matrix(changedf$ha,
                    nrow = 7,
                    ncol = 7)
rownames(changemat) <- shortnames
colnames(changemat) <- shortnames
```

TABLE 7.2

Change matrix from 2001–2019 in Walton County expressed as a percent of the area of each 2001 class, with 2001 classes are rows and 2019 classes as columns.

	Wat	Dev	Bare	For	Grass	Crop	Wet
Wat	94.5	0.6	0.0	1.8	0.7	0.0	2.4
Dev	0.0	100.0	0.0	0.0	0.0	0.0	0.0
Bare	1.3	38.9	48.3	8.5	3.0	0.0	0.1
For	0.4	3.9	0.1	89.5	6.0	0.0	0.0
Grass	0.1	5.2	0.2	12.7	81.6	0.2	0.0
Crop	0.0	7.5	0.0	0.0	0.0	92.4	0.0
Wet	1.1	0.2	0.0	0.0	0.1	0.0	98.6

In this example, the 2001 classes are displayed as rows, and the 2019 classes are displayed as columns. For a particular class in 2001 (row), the numbers in each column show how much area of that class has transitioned to the other land cover classes.

When interpreting a change matrix, note that the numbers along the diagonal from upper left to lower right represent cells that did not change, including water cells that remained water, developed cells that remained developed, etc. Calculating the total of each row will provide the area of each class in 2001, and the total of each column will provide the area of each class in 2019.

It is also useful to generate the transition matrix with the percentage of each 2001 class that changed to each 2019 class (Table 7.2).

```
percmat <- matrix(changedf$perc,
                  nrow = 7,
                  ncol = 7)
rownames(percmat) <- shortnames
colnames(percmat) <- shortnames
```

Using the original data frame, the data in the change matrix can be plotted as a grouped bar chart. Each group represents a 2001 land cover class, and each bar within a group represents a 2019 land cover class (Figure 7.13).

```
ggplot(data = changedf) +
  geom_bar(aes(x = X2001,
               y = ha,
               group = X2019,
               fill = X2019),
```

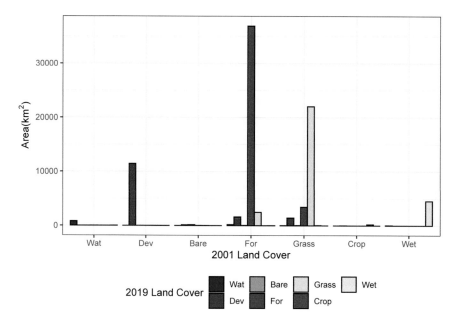

FIGURE 7.13
Area of each 2001 land cover class changing to each 2019 land cover class.

```
            color = "black",
            position = "dodge",
            stat = "identity") +
scale_fill_manual(name = "2019 Land Cover", values = newcols) +
labs(x = "2001 Land Cover",
     y = expression("Area(km"^2*")")) +
theme_bw() +
theme(legend.position="bottom")
```

Plotting change as percent area (relative to the 2001 area of each land cover class) makes it easier to see the changes occurring in the less abundant classes (Figure 7.14).

```
ggplot(data = changedf) +
  geom_bar(aes(x = X2001,
               y = perc,
               group = X2019,
               fill = X2019),
           color = "black",
```

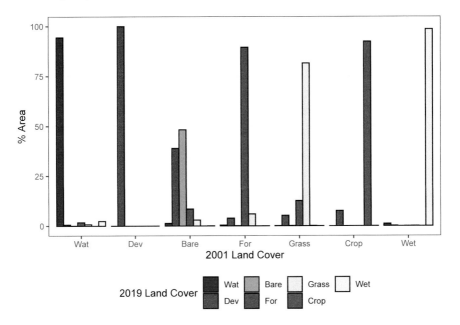

FIGURE 7.14
Percentage of each 2001 land cover class changing to each 2019 land cover class.

```
        position = "dodge",
        stat = "identity") +
scale_fill_manual(name = "2019 Land Cover", values = newcols) +
labs(x = "2001 Land Cover", y = "% Area") +
theme_bw() +
theme(legend.position="bottom")
```

7.6   Mapping Specific Land Cover Changes

To better see where specific types of change are occurring, new classified data characterizing change events can be generated. The example will focus on changes in forest cover. Forest loss is classified as pixels with forest in 2001 and no forest in 2019, and forest gain is classified as pixels with no forest in 2001 and forest in 2019.

```
nlcd01_rc <- nlcd_rc[["2001"]]
nlcd19_rc <- nlcd_rc[["2019"]]
forloss <- nlcd01_rc == 4 & nlcd19_rc != 4
forgain <- nlcd01_rc != 4 & nlcd19_rc == 4
```

The `forloss` and `forgain` rasters contain binary variables that indicate where forest loss and forest gain occurred. Using a mathematical expression, they are combined into a single layer where $1 = $ no forest change, $2 = $ forest loss, and $3 = $ forest gain.

```
forchange <- 1 + forloss + forgain * 2
```

This raster can be used to visualize the locations where forest cover has increased and decreased. Forests have been mostly lost in the northwest corner of the study area and mostly gained in the southeast corner, with a mixture of loss and gain in most other locations (Figure 7.15).

```
forchange_df <- rasterdf(forchange)

names_chg <- c("No Change",
               "Forest Loss",
               "Forest Gain")
cols_chg <- c("lightgrey",
              "red",
              "blue")

ggplot(forchange_df) +
  geom_raster(aes(x = x,
                  y = y,
                  fill = as.character(value))) +
  scale_fill_manual(name = "Land cover",
                    values = cols_chg,
                    labels = names_chg,
                    na.translate = FALSE) +
  coord_sf(expand = FALSE) +
  theme_void()
```

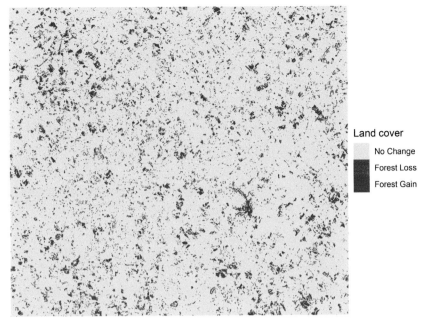

FIGURE 7.15
Forest loss and gain between 2001 and 2019.

7.7 Practice

1. Conduct focal analyses of forest land cover using 100 m, 500 m, 1000 m, and 2000 m circular window sizes. Display the maps of the results using a faceted plot.

2. Generate two Walton County change matrices, one from 2001–2008 and another from 2011–2019. Compare these results to assess whether land cover transitions have been consistent over time.

3. Generate a classified change map highlighting areas of gain and loss of the Grass/Shrub land cover class.

8

Coordinate Reference Systems

All geospatial data have a coordinate reference system (CRS) that defines how the two-dimensional coordinates used for mapping and analysis are related to real locations on the Earth's surface. Important components of the CRS include the ellipsoid that models the shape of the Earth and the datum that connects the ellipsoid to the surface of the earth. The datum and ellipsoid define a geographic coordinate system, where coordinates of latitude and longitude are measured as angles from the center of the ellipsoid to a point on the Earth's surface. To define a two-dimensional Cartesian coordinate system, a map projection must be applied to translate the curved surface of the Earth onto a two-dimensional plane. Commonly used projections include Transverse Mercator, Albers Equal Area, and Lambert Conformal Conic. Additional parameters must be supplied to define the projection, including the distance unit, central meridian, central parallel, standard parallels, and false eastings and northings. Coordinate systems like Universal Transverse Mercator offer predefined projections for a set of zones distributed across the globe, allowing users to easily select a local projection without having to specify all the parameters.

Frequently, it is necessary to combine multiple sources of data with different projections before conducting an analysis. The R packages for geospatial data processing all contain functions for defining projections and reprojecting data. It is usually advisable to do any necessary reprojection within your R script rather than reprojecting the data externally and then importing it into R. Combining all of the data processing steps within a single script makes it easier to replicate the workflow and is more straightforward and convenient than using multiple software tools.

```
library(tidyverse)
library(sf)
library(terra)
library(colorspace)
```

Here, the entire collection of tidyverse packages is loaded with library(tidyverse). This approach will install **ggplot**, **dplyr**, **tidyr**, **readr**, and several other core tidyverse packages. Before you use this shortcut, be

DOI: 10.1201/9781003326199-8   151 sure to run `install.packages("tidyverse")` to ensure that all of the packages are installed.

8.1   Reprojecting Vector Data

These examples will use a dataset of U.S. county boundaries from the U.S. Census Bureau. To keep things simple, Alaska, Hawaii, and the territories are removed so that the data only include the 48 conterminous states. The `st_read()` import function reads the state FIPS code, `STATEFP`, to a factor. To convert it back to a number, it must first be converted to a character and then to a numeric variable. After this conversion, it is easier to filter the data to the desired states.

```
county <- st_read("cb_2018_us_county_20m.shp", quiet = TRUE)
county <- county %>%
  mutate(state = as.numeric(as.character(STATEFP))) %>%
  filter(state != 2, state != 15, state < 60)
```

To find out what projection the data are in, the `st_crs()` function in the **sf** package can be used. The data are in geographic coordinates (longitude and latitude) and are referenced to the GRS80 ellipsoid and NAD83 datum. The projection also has a European Petroleum Survey Group (EPSG) code. The EPSG maintains a standard list of codes for geographic objects, including coordinate systems, datums, and spheroids. In this case, the spatial reference identifier (SRID) for this geographic coordinate system is 4269. These codes are handy because when reprojecting data, it is easier to just enter a single code than all the detailed projection parameters shown in the printed output from `st_crs()`.

```
st_crs(county)
## Coordinate Reference System:
##    User input: NAD83
##    wkt:
## GEOGCRS["NAD83",
##     DATUM["North American Datum 1983",
##         ELLIPSOID["GRS 1980",6378137,298.257222101,
##             LENGTHUNIT["metre",1]]],
##     PRIMEM["Greenwich",0,
##         ANGLEUNIT["degree",0.0174532925199433]],
##     CS[ellipsoidal,2],
##         AXIS["latitude",north,
```

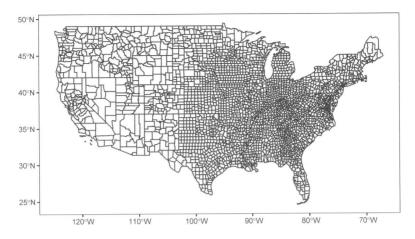

FIGURE 8.1
U.S. counties mapped in a geographic coordinate system.

```
##              ORDER[1],
##              ANGLEUNIT["degree",0.0174532925199433]],
##          AXIS["longitude",east,
##              ORDER[2],
##              ANGLEUNIT["degree",0.0174532925199433]],
##      ID["EPSG",4269]]
```

Because longitude and latitude are angles and not cartesian coordinates, simply plotting them on a map results in a distorted view of the United States. The states appear stretched out in an East-West direction, particularly in the north, where one unit of longitude covers a longer distance than in the south (Figure 8.1).

```
ggplot(data = county) +
  geom_sf(fill = NA) +
  theme_bw()
```

The st_crs() function returns a list containing various data elements that describe the projection, and the projection name, EPSG code, PROJ.4 string code, and WKT string code can be viewed by selecting the appropriate list elements. The following code extracts and prints the projection name and EPSG code.

```
st_crs(county)$Name
## [1] "NAD83"
```

```
st_crs(county)$epsg
## [1] 4269
```

The PROJ.4 format, stored in the proj4string element, is an older, abbreviated format for storing CRS parameters. This format is no longer recommended for defining coordinate systems in R because of changes in the underlying RPROJ software library.

```
st_crs(county)$proj4string
## [1] "+proj=longlat +datum=NAD83 +no_defs"
```

Well-known text (WKT) is a more comprehensive format that is currently recommended for defining and storing CRS information in R. However, WKT can be difficult for humans to read. Extracting the WktPretty element and printing it to the console using the writeLines() function provide a formatted version of the WKT string that makes it easier to see the various CRS parameters.

```
writeLines(st_crs(county)$WktPretty)
## GEOGCS["NAD83",
##     DATUM["North_American_Datum_1983",
##         SPHEROID["GRS 1980",6378137,298.257222101]],
##     PRIMEM["Greenwich",0],
##     UNIT["degree",0.0174532925199433,
##         AUTHORITY["EPSG","9122"]],
##     AXIS["Latitude",NORTH],
##     AXIS["Longitude",EAST],
##     AUTHORITY["EPSG","4269"]]
```

The st_transform() function is used to reproject **sf** objects. The first argument is the spatial dataset to reproject. The second argument is the coordinate system into which the data will be reprojected. The coordinate system information can be entered in several different ways. Here, we simply enter another EPSG code (5070) which is a commonly-used Albers Equal Area projection for the conterminous United States.

```
county_aea <- st_transform(county, 5070)
writeLines(st_crs(county_aea)$WktPretty)
## PROJCS["NAD83 / Conus Albers",
##     GEOGCS["NAD83",
##         DATUM["North_American_Datum_1983",
##             SPHEROID["GRS 1980",6378137,298.257222101]],
##         PRIMEM["Greenwich",0],
```

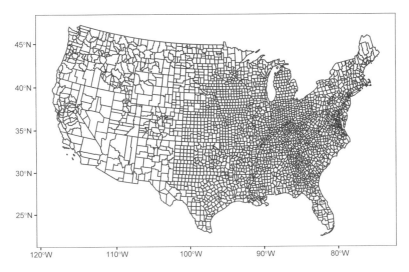

FIGURE 8.2
U.S. counties projected into an Albers Equal Area coordinate system for the
conterminous United States.

```
##              UNIT["degree",0.0174532925199433,
##                  AUTHORITY["EPSG","9122"]],
##              AUTHORITY["EPSG","4269"]],
##          PROJECTION["Albers_Conic_Equal_Area"],
##          PARAMETER["latitude_of_center",23],
##          PARAMETER["longitude_of_center",-96],
##          PARAMETER["standard_parallel_1",29.5],
##          PARAMETER["standard_parallel_2",45.5],
##          PARAMETER["false_easting",0],
##          PARAMETER["false_northing",0],
##          UNIT["metre",1],
##          AXIS["Easting",EAST],
##          AXIS["Northing",NORTH],
##          AUTHORITY["EPSG","5070"]]
```

In the resulting map, note that the x and y axes of the coordinate system no
longer correspond to the graticules of latitude and longitude (Figure 8.2).

```
ggplot(data = county_aea) +
  geom_sf(fill = NA) +
  theme_bw()
```

Using EPSG codes is the recommended method for specifying coordinate systems because specifying these codes is much simpler than entering all the underlying projection details. The following websites can be used to look up EPSG codes for different coordinate reference systems.

- https://spatialreference.org/
- https://epsg.org/home.html

Another way to define the projection is to enter all of the projection details directly using the well-known text (WKT version 2) projection format. This approach requires the user to provide a string containing the specification and parameters of the datum and coordinate system. This example specifies a "bad" projection that is based on the Albers projection for the conterminous United States. In this example, the standard parallels have been shifted northward to the Artic Circle, which should result in considerable distortion.

```
badcrs_wkt <-
'PROJCS["BadAlbers",
GEOGCS["NAD83",
  DATUM["North_American_Datum_1983",
    SPHEROID["GRS 1980",6378137,298.257222101]],
  PRIMEM["Greenwich",0],
  UNIT["degree",0.0174532925199433]],
PROJECTION["Albers_Conic_Equal_Area"],
PARAMETER["latitude_of_center",37.5],
PARAMETER["longitude_of_center",-96],
PARAMETER["standard_parallel_1",75],
PARAMETER["standard_parallel_2",80],
PARAMETER["false_easting",0],
PARAMETER["false_northing",0],
UNIT["metre",1],
AXIS["Easting",EAST],
AXIS["Northing",NORTH]]'

mybadcrs <- st_crs(badcrs_wkt)
county_bad <- st_transform(county, mybadcrs)
```

That resulting map does not look very good. There is an unnecessary distortion because the standard parallels are outside of the area being mapped (Figure 8.3).

```
ggplot(data = county_bad) +
  geom_sf(fill = NA) +
  theme_bw()
```

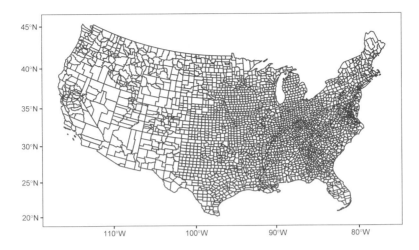

FIGURE 8.3
U.S. counties projected into an Albers Equal Area coordinate system with
standard parallels located too far north.

Often, the most straightforward way to specify a CRS is to use the information
from another geospatial dataset that is in the desired CRS. An easy way to do
this is to use the st_crs() function to extract the coordinate reference system
from the other spatial dataset. Here, we reproject the county dataset with the
"bad" projection back to the Albers Equal Area projection from the county_aea
dataset (Figure 8.4).

```
county_fixed <- st_transform(county_bad, st_crs(county_aea))
ggplot(data = county_fixed) +
  geom_sf(fill = NA) +
  theme_bw()
```

This approach is a straightforward way to make sure all the data are in the same
projection. Start by projecting one of your datasets into the desired coordinate
system and then project all the other datasets to match this baseline dataset.

Here are a few more examples looking at just the state of New York. It looks
a bit tilted because it is at the edge of the conterminous U.S. and thus gets
distorted by the national-level Albers projection (Figure 8.5).

```
newyork <- filter(county_aea, state == 36)
ggplot(data = newyork) +
  geom_sf(fill = NA) +
  theme_bw()
```

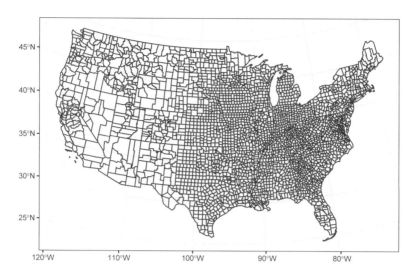

FIGURE 8.4
U.S. counties projected back to an Albers Equal Area coordinate system for
the conterminous United States.

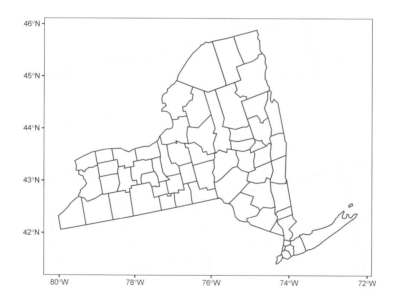

FIGURE 8.5
New York counties in an Albers Equal Area coordinate system for the conterminous United States.

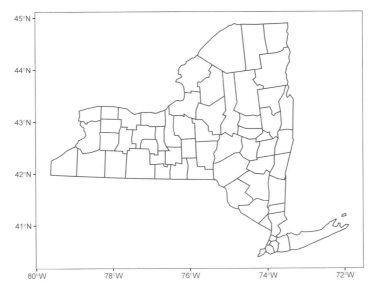

FIGURE 8.6
New York counties in the UTM zone 18 north coordinate system.

It can be made to look a bit better by putting it in a more localized UTM projection, Universal Transverse Mercator (UTM) zone 18 north. In this case, the EPSG code of 26918 represents UTM zone 18 north with a NAD83 datum (Figure 8.6).

```
ny_utm <- st_transform(newyork, 26918)
writeLines(st_crs(ny_utm)$WktPretty)
## PROJCS["NAD83 / UTM zone 18N",
##     GEOGCS["NAD83",
##         DATUM["North_American_Datum_1983",
##             SPHEROID["GRS 1980",6378137,298.257222101]],
##         PRIMEM["Greenwich",0],
##         UNIT["degree",0.0174532925199433,
##             AUTHORITY["EPSG","9122"]],
##         AUTHORITY["EPSG","4269"]],
##     PROJECTION["Transverse_Mercator"],
##     PARAMETER["latitude_of_origin",0],
##     PARAMETER["central_meridian",-75],
##     PARAMETER["scale_factor",0.9996],
##     PARAMETER["false_easting",500000],
##     PARAMETER["false_northing",0],
##     UNIT["metre",1],
```

```
##       AXIS["Easting",EAST],
##       AXIS["Northing",NORTH],
##       AUTHORITY["EPSG","26918"]]

ggplot(data = ny_utm) +
  geom_sf(fill = NA) +
  theme_bw()
```

8.2   Reprojecting Raster Data

Here, we will take another look at the Walton County raster dataset that was used in Chapter 7. The raster grid is 1579 rows by 1809 columns, and each cell is a 30 m square.

```
landcov <- rast("NLCD_2016_Land_Cover_Walton.tiff")
nrow(landcov)
## [1] 1579
ncol(landcov)
## [1] 1809
res(landcov)
## [1] 30 30
ext(landcov)[1:4]
##     xmin     xmax     ymin     ymax
## 1096815 1151085 1237395 1284765
```

The **terra** function `crs()` returns detailed projection information for Spa-tRaster objects. The `writeLines()` function parses the projection parameters and writes them to the console in a more readable format. These data have a similar Albers Equal Area projection to the one used for the county data.

```
writeLines(crs(landcov))
## PROJCRS["Albers_Conical_Equal_Area",
##     BASEGEOGCRS["NAD83",
##         DATUM["North American Datum 1983",
##             ELLIPSOID["GRS 1980",6378137,298.257222101004,
##                 LENGTHUNIT["metre",1]]],
##         PRIMEM["Greenwich",0,
##             ANGLEUNIT["degree",0.0174532925199433]],
##         ID["EPSG",4269]],
##     CONVERSION["Albers Equal Area",
```

```
##          METHOD["Albers Equal Area",
##              ID["EPSG",9822]],
##          PARAMETER["Latitude of false origin",23,
##              ANGLEUNIT["degree",0.0174532925199433],
##              ID["EPSG",8821]],
##          PARAMETER["Longitude of false origin",-96,
##              ANGLEUNIT["degree",0.0174532925199433],
##              ID["EPSG",8822]],
##          PARAMETER["Latitude of 1st standard parallel",29.5,
##              ANGLEUNIT["degree",0.0174532925199433],
##              ID["EPSG",8823]],
##          PARAMETER["Latitude of 2nd standard parallel",45.5,
##              ANGLEUNIT["degree",0.0174532925199433],
##              ID["EPSG",8824]],
##          PARAMETER["Easting at false origin",0,
##              LENGTHUNIT["metre",1],
##              ID["EPSG",8826]],
##          PARAMETER["Northing at false origin",0,
##              LENGTHUNIT["metre",1],
##              ID["EPSG",8827]]],
##      CS[Cartesian,2],
##          AXIS["easting",east,
##              ORDER[1],
##              LENGTHUNIT["metre",1,
##                  ID["EPSG",9001]]],
##          AXIS["northing",north,
##              ORDER[2],
##              LENGTHUNIT["metre",1,
##                  ID["EPSG",9001]]]]
```

To see only the projection name, use the describe = TRUE argument with crs()
and extract the name element from the data frame that is returned.

```
crs(landcov, describe = TRUE)$name
## [1] "Albers_Conical_Equal_Area"
```

The st_crs() function from the **sf** package can also be used to extract CRS
information from spatRaster objects. This approach prints the CRS in a format
that is a bit more compact and easy to read than the output of the crs()
function.

```
writeLines(st_crs(landcov)$WktPretty)
## PROJCS["Albers_Conical_Equal_Area",
```

```
##      GEOGCS["NAD83",
##          DATUM["North_American_Datum_1983",
##              SPHEROID["GRS 1980",6378137,298.257222101004]],
##          PRIMEM["Greenwich",0],
##          UNIT["degree",0.0174532925199433,
##              AUTHORITY["EPSG","9122"]],
##          AUTHORITY["EPSG","4269"]],
##      PROJECTION["Albers_Conic_Equal_Area"],
##      PARAMETER["latitude_of_center",23],
##      PARAMETER["longitude_of_center",-96],
##      PARAMETER["standard_parallel_1",29.5],
##      PARAMETER["standard_parallel_2",45.5],
##      PARAMETER["false_easting",0],
##      PARAMETER["false_northing",0],
##      UNIT["metre",1,
##          AUTHORITY["EPSG","9001"]],
##      AXIS["Easting",EAST],
##      AXIS["Northing",NORTH]]
```

Before mapping the land cover data, they are reclassified into seven broader classes as in Chapter 7.

```
oldclas <- unique(landcov)
newclas <- c(1, 2, 2, 2, 2, 3, 4, 4, 4, 5, 5, 5, 6, 7, 7)
lookup <- data.frame(oldclas, newclas)
landcov_rc <- classify(landcov, lookup)
newnames <- c("Water",
              "Developed",
              "Barren",
              "Forest",
              "GrassShrub",
              "Cropland",
              "Wetland")
newcols <- c("mediumblue",
             "red2",
             "gray60",
             "darkgreen",
             "yellow2",
             "orange4",
             "paleturquoise2")
newcols2 <- desaturate(newcols, amount = 0.3)
```

The data are then converted to a data frame using the `rasterdf()` function and mapped (Figure 8.7).

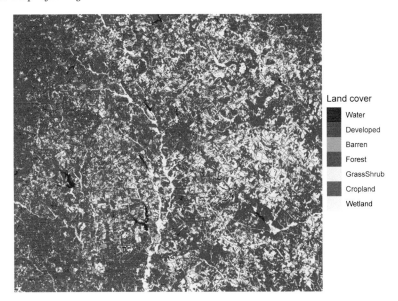

FIGURE 8.7
Walton County land cover data in an Albers Equal Area coordinate reference
system for the conterminous United States.

```
source("rasterdf.R")
landcov_df <- rasterdf(landcov_rc)

ggplot(data = landcov_df) +
  geom_raster(aes(x = x,
                  y = y,
                  fill = as.character(value))) +
  scale_fill_manual(name = "Land cover",
                    values = newcols2,
                    labels = newnames,
                    na.translate = FALSE) +
  coord_sf(expand = FALSE) +
  theme_void()
```

These data will be projected into a UTM projection (zone 17 north) and
NAD 83 datum using an EPSG code (26917) as was done for the New York
dataset. Raster data are reprojected using the `project()` function from the
terra package. When raster data are reprojected, the values must be resampled
onto a new raster grid in the updated projection. Like the `st_transform()`
function, the first argument is the dataset to be reprojected, and the second
argument is the new CRS. Note that the format for specifying EPSG codes is

a bit different when working with rasters in the **terra** package. Instead of just a number, a string must be provided with `epsg:` preceding the number.

By default, the `project()` function uses bilinear interpolation, which is appropriate for continuous data such as elevation or temperature. For land cover data, interpolated values that fall in between the numerical class codes would be meaningless. Therefore, the argument `method = "near"` must be specified to use nearest neighbor interpolation, which preserves the class codes. For continuous data, this argument can be omitted or `method = "bilinear"` can be explicitly used. Specifying `res = 30` ensures that the output raster has the same cell size (30 m) as the input raster.

```
landcov_utm <- project(landcov_rc, "epsg:26917",
                       method = "near",
                       res = 30)
crs(landcov_utm, describe = TRUE)$name
## [1] "NAD83 / UTM zone 17N"
```

The coordinate reference system has been changed, and the numbers of rows and columns in the raster have increased as well.

```
nrow(landcov_utm)
## [1] 1828
ncol(landcov_utm)
## [1] 2047
res(landcov_utm)
## [1] 30 30
```

These changes are apparent in the map of the reprojected Walton County data (Figure 8.8). As was observed with New York State, the orientation of the dataset has shifted after projecting into UTM. Because raster layers must be rectangular grids, the overall size of the raster is increased to accommodate the new orientation of the dataset, and NA values are used to account for cells in the new raster that are outside of the boundaries of the original dataset.

```
landcovutm_df <- rasterdf(landcov_utm)
ggplot(data = landcovutm_df) +
  geom_raster(aes(x = x,
                  y = y,
                  fill = as.character(value))) +
  scale_fill_manual(name = "Land cover",
                    values = newcols2,
                    labels = newnames,
                    na.translate = FALSE) +
```

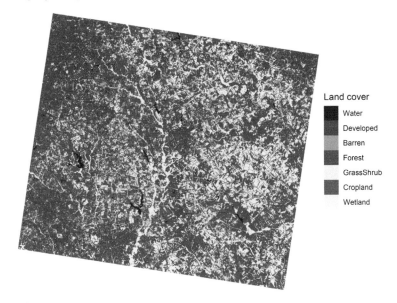

FIGURE 8.8
Walton County land cover data in Universal Transverse Mercator coordinate
system.

```
coord_sf(expand = FALSE) +
theme_void()
```

As with vector data, a simple way to reproject raster data is to specify a raster
dataset from which to obtain information about the desired CRS, cell size, grid
size, and grid origin. In this example, the Walton County data are reprojected
from UTM back to the original Albers Equal Area coordinate system. The
first argument to the `project()` function is the raster dataset to be projected,
and the second argument is another raster dataset that provides the CRS and
raster grid information.

```
landcov_goback <- project(landcov_utm,
                          landcov,
                          method = "near")
```

Now the `landcov_goback` dataset has the same projection as the original `landcov`
dataset. The `compareGeom()` function can be used to confirm that the two rasters
have the exact same geometry, with identical extents, numbers of rows and
columns, coordinate reference systems, cell sizes, and origins.

```
crs(landcov_goback, describe = TRUE)$name
## [1] "Albers_Conical_Equal_Area"
crs(landcov_rc, describe = TRUE)$name
## [1] "Albers_Conical_Equal_Area"
compareGeom(landcov_rc, landcov_goback)
## [1] TRUE
```

The landcov and landcov_goback rasters have the same geometry, but what about the actual data? Here, the freq() function is used here to compare the number of cells in each land cover class to see if projecting into UTM and then back to the original Albers Equal Area coordinate system has altered the data.

```
freq(landcov_rc)
##        layer value    count
## [1,]      1     1    37165
## [2,]      1     2   687865
## [3,]      1     3     7928
## [4,]      1     4  1296342
## [5,]      1     5   691107
## [6,]      1     6     6629
## [7,]      1     7   129375
freq(landcov_goback)
##        layer value    count
## [1,]      1     1    37182
## [2,]      1     2   687909
## [3,]      1     3     7916
## [4,]      1     4  1296069
## [5,]      1     5   691183
## [6,]      1     6     6631
## [7,]      1     7   129411
```

It appears that the distributions of land cover types do not match exactly. To look at the changes more closely, the crosstab() function is applied in a similar manner to the way it was used to generate change matrices in Chapter 7. Here, instead of looking at change over time, the crosstabulation quantifies differences between landcov_rc and landcov_goback .

```
projcomp <- crosstab(c(landcov_rc, landcov_goback))
shortnames <- c("Wat",
                "Dev",
                "Bare",
                "For",
                "Grass",
```

TABLE 8.1

Crosstabulation of cell values from the original Walton County land cover dataset (rows) and the same data projected into UTM and then back to the original Albers Equal Area CRS.

	Wat	Dev	Bare	For	Grass	Crop	Wet
Wat	36508	72	7	377	157	0	43
Dev	61	678023	76	5156	4370	14	137
Bare	7	70	7678	57	113	0	3
For	408	5299	59	1284073	5101	5	1348
Grass	166	4297	93	5081	681232	44	164
Crop	0	10	1	6	44	6567	1
Wet	32	138	2	1319	166	1	127715

```
                "Crop",
                "Wet")
rownames(projcomp) <- shortnames
colnames(projcomp) <- shortnames
```

When examining the crosstabulation, the numbers along the diagonal from upper left to lower right are counts of cells that have the same land cover code in both datasets (8.1). Numbers in the upper right and lower left off-diagonal triangles indicate cells that have different values in the landcov_rc and landcov_goback datasets. It is clear from these results that while most of the cell values are the same in both datasets, reprojecting the landcov_goback dataset twice has altered the land cover codes in some of the cells.

This is an important point when working with raster data. Reprojecting rasters always changes the data because they have to be resampled to a new grid in the new coordinate system. This process alters the distribution of cell-level data values and is not exactly reversible. The bottom line is to avoid reprojecting raster datasets more than once. Instead, go back to the original data and reproject from the native coordinate system if you can. When combining raster and vector data, avoid reprojecting the rasters if possible and instead reproject the vector data to match the CRS of the rasters.

8.3 Specifying Coordinate Reference Systems

All spatial data must use some type of coordinate reference system to specify locations. However, in some cases, a spatial dataset may lack information about

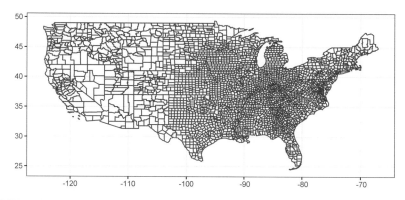

FIGURE 8.9
U.S. counties mapped in a geographic coordinate system using a dataset that
is missing CRS information.

its coordinate reference system. For example, the PRJ file of a shapefile may
get lost when copying the data from one location to another. In this case, it is
often necessary to do some detective work to determine the actual coordinate
reference system and then specify it after the data have been imported into R.
This example reads in another version of the U.S. county shapefile where the
PRJ file has been deleted.

```
county_nocrs <- st_read("cb_2018_us_county_noPRJ.shp", quiet = TRUE)
county_nocrs <- county_nocrs %>%
  mutate(state = as.numeric(as.character(STATEFP))) %>%
  filter(state != 2, state != 15, state < 60)
```

These data can still be mapped (Figure 8.9), but the software does not know
whether the coordinates represent latitude and longitude or some type of
distance units relative to the origin of a projected CRS. This is similar to a
situation where there is a dataset with nonspatial measurements such as tree
heights, but the units of measurement have not been recorded. We cannot
make use of the tree heights until we find out if they are in centimeters, inches,
feet, or meters. Similarly, geospatial data cannot be used effectively without
knowledge of the underlying CRS.

```
st_crs(county_nocrs)
## Coordinate Reference System: NA
ggplot(data = county_nocrs) +
  geom_sf(fill = NA) +
  theme_bw()
```

If the current CRS is not defined, it is not possible to reproject the data into a new CRS. Attempting to use the st_transform() function with the county_nocrs data will return an error. Referring back to the tree height example, we cannot convert tree heights with unknown units to meters until we know if the current units are centimeters, inches, or feet, or perhaps they are already in meters.

```
county_newcrs <- st_transform(county_nocrs, 4269)
```

Although the coordinate system is undefined, a look at the map clearly shows that the data have geographic coordinates (Figure 8.9). This is apparent from the range of coordinate values as well as the characteristic distortion that we see when angular coordinates are plotted on a two-dimensional map. The problem is that the necessary metadata to identify the coordinate system is not present. To specify a missing coordinate system, the st_set_crs() function is used with the EPSG code for geographic coordinates in a NAD83 datum (4269).

```
county_crs <- st_set_crs(county_nocrs, 4269)
writeLines(st_crs(county_crs)$WktPretty)
## GEOGCS["NAD83",
##     DATUM["North_American_Datum_1983",
##         SPHEROID["GRS 1980",6378137,298.257222101]],
##     PRIMEM["Greenwich",0],
##     UNIT["degree",0.0174532925199433,
##         AUTHORITY["EPSG","9122"]],
##     AXIS["Latitude",NORTH],
##     AXIS["Longitude",EAST],
##     AUTHORITY["EPSG","4269"]]
```

Now that the proper coordinate reference system has been assigned, it is possible to project the data into another CRS.

```
county_newcrs <- st_transform(county_crs, 5070)
writeLines(st_crs(county_newcrs)$WktPretty)
## PROJCS["NAD83 / Conus Albers",
##     GEOGCS["NAD83",
##         DATUM["North_American_Datum_1983",
##             SPHEROID["GRS 1980",6378137,298.257222101]],
##         PRIMEM["Greenwich",0],
##         UNIT["degree",0.0174532925199433,
##             AUTHORITY["EPSG","9122"]],
##         AUTHORITY["EPSG","4269"]],
##     PROJECTION["Albers_Conic_Equal_Area"],
```

```
##      PARAMETER["latitude_of_center",23],
##      PARAMETER["longitude_of_center",-96],
##      PARAMETER["standard_parallel_1",29.5],
##      PARAMETER["standard_parallel_2",45.5],
##      PARAMETER["false_easting",0],
##      PARAMETER["false_northing",0],
##      UNIT["metre",1],
##      AXIS["Easting",EAST],
##      AXIS["Northing",NORTH],
##      AUTHORITY["EPSG","5070"]]
```

It is also possible to assign coordinate reference systems to raster data using the `crs()` function in the `terra()` package. For example, if `myraster_nocrs` is a raster dataset with no CRS specified, it can be assigned a coordinate system by specifying an EPSG code or assigning the CRS from another raster dataset.

```
crs(raster_nocrs) <- "epsg:4326"
crs(raster_nocrs) <- crs(raster_withcrs)
```

This chapter has provided a very basic introduction to working with coordinate references systems in R, with a focus on just a few examples from North America. For more background on the topic, the book *GIS Fundamentals* (Bolstad, 2019) is an excellent reference. *Geocomputation with R* (Lovelace et al., 2019) also contains a chapter on reprojecting geographic data in R.

8.4   Practice

1. Generate an `sf` object for the state of California (state FIPS code = 6). Compare maps of the state in a geographic projection (EPSG code 4269), Albers equal area projection (EPSG code 5070), and UTM zones 10 and 11 (EPSG codes 26910 and 26911). Explore how the different coordinate reference systems affect the shape and orientation of the state.

2. Reproject the Walton County land cover dataset into the same geographic coordinate system used in the U.S. county dataset. Specify a resolution of 0.008333 degrees (30 arc seconds) for the reprojected raster.

9

Combining Vector Data with Continuous Raster Data

To conduct geospatial analyses, it is usually necessary to combine data from multiple sources. Often, this involves using vector data as well as raster data. For example, geospatial climate and weather data are typically provided as rasters, while geospatial data on the demographics of human populations are typically referenced to polygons. To understand how different populations subgroups may be exposed to different climate and meteorological hazards, it is necessary to integrate data with fundamentally different geospatial structures.

The next two chapters will provide several examples of how vector and raster data can be combined in R to carry out common types of analyses. This chapter will present several analyses using the PRISM dataset (`http://www.prism.oregonstate.edu`). This is a widely-used interpolated climate dataset for the United States that is based on data from meteorological stations combined with other ancillary datasets such as elevation (Daly et al., 2002).

```
library(tidyverse)
library(sf)
library(terra)
library(prism)
library(tigris)
```

9.1   Accessing Data with R Packages

In previous chapters, various functions have been used to import data into R and convert them into appropriate objects for tabular data and geospatial data in vector and raster formats. This chapter will use a different approach for bringing data into R. County boundaries and gridded climate and meteorological data will be imported directly from online archives using two R packages specifically designed to facilitate data access: **tigris** (Walker, 2022) and **prism** (Edmund and Bell, 2020).

DOI: 10.1201/9781003326199-9

To start, we will use the **tigris** package to download a county shapefile similar to the one that was used in Chapter 8 on coordinate reference systems. The county() function is used with the cb = TRUE argument to indicate that a dataset with generalized county boundaries will be downloaded and resolution = '20m' to indicate the scale (1:20 million) of the data. The function returns an sf object that contains columns with state and county names, codes, and areas along with geometry information for each county.

```
county <- counties(cb = TRUE,
                   resolution = '20m')
class(county)
glimpse(county)
```

The STATEFP and GEOID fields are initially read in as characters, but they are converted to numbers to make it easier to join with the zonal summary table later on. Then, as in Chapter 8, the dataset is filtered down to only the 48 conterminous United States.

```
county <- county %>%
  mutate(state = as.numeric(STATEFP),
         fips = as.numeric(GEOID)) %>%
  filter(state != 2, state != 15, state < 60)
```

As discussed in Chapter 8, it is important to know the coordinate reference system of the data. These county data are in a geographic coordinate system (longitude and latitude) with a NAD83 datum.

```
writeLines(st_crs(county)$WktPretty)
## GEOGCS["NAD83",
##     DATUM["North_American_Datum_1983",
##         SPHEROID["GRS 1980",6378137,298.257222101]],
##     PRIMEM["Greenwich",0],
##     UNIT["degree",0.0174532925199433,
##         AUTHORITY["EPSG","9122"]],
##     AXIS["Latitude",NORTH],
##     AXIS["Longitude",EAST],
##     AUTHORITY["EPSG","4269"]]
```

The **prism** package automates data downloading and importing of PRISM climate data. Setting options(prism.path = ".") downloads and stores the data in the current working directly. To download the data to a different location in the file system, a directory path can be specified here instead. The code in this example downloads the 30-year climatology of annual precipitation for the conterminous U.S. using the get_prism_normals() function. The `type = 'ppt'` argument specifies that precipitation data will be downloaded. Other types of data that can be downloaded are shown on the help pages for `get_prism_normals()` and other **prism** functions. The `resolution` argument can be used to select either `4km` or `800m` grids. The `annual = TRUE` argument selects annual instead of monthly summaries, and the `keepZip = TRUE` argument saves the downloaded ZIP archives.

```
options(prism.path = ".")
get_prism_normals(type = 'ppt',
                  resolution = '4km',
                  annual = T,
                  keepZip = TRUE)
```

The `get_prism_monthlys()` function can similarly be used to download monthly meteorological summaries for a specified range of years and months.

```
get_prism_monthlys(type = 'ppt',
                   years = 2018,
                   mon=1:12,
                   keepZip = TRUE)
```

The **prism** package also has functions for managing these downloaded data. The `prism_archive_ls()` function produces a vector containing the names of all downloaded PRISM datasets.

```
prism_files <- prism_archive_ls()
prism_files
##  [1] "PRISM_ppt_30yr_normal_4kmM3_annual_bil"
##  [2] "PRISM_ppt_stable_4kmM3_201801_bil"
##  [3] "PRISM_ppt_stable_4kmM3_201802_bil"
##  [4] "PRISM_ppt_stable_4kmM3_201803_bil"
##  [5] "PRISM_ppt_stable_4kmM3_201804_bil"
##  [6] "PRISM_ppt_stable_4kmM3_201805_bil"
##  [7] "PRISM_ppt_stable_4kmM3_201806_bil"
##  [8] "PRISM_ppt_stable_4kmM3_201807_bil"
##  [9] "PRISM_ppt_stable_4kmM3_201808_bil"
## [10] "PRISM_ppt_stable_4kmM3_201809_bil"
## [11] "PRISM_ppt_stable_4kmM3_201810_bil"
## [12] "PRISM_ppt_stable_4kmM3_201811_bil"
## [13] "PRISM_ppt_stable_4kmM3_201812_bil"
```

Each PRISM dataset consists of multiple files and is stored in a separate folder, which conveniently has the exact same name as the dataset. The files include a bit interleaved by line (BIL) image as well as other auxiliary files containing

projection and header information along with metadata. The following code
generates a vector of paths to the BIL files. The "." at the beginning of the path
indicates that the folder location is relative to the user's working directory.
The rast() function is used to create a SpatRaster object containing the
30-year precipitation normals (the first element in the prism_paths vector)
and another containing monthly precipitation rasters for 2018 (elements two
through thirteen in the prism_paths vector).

```
prism_paths <- file.path(".",
                    prism_files,
                    paste0(prism_files, ".bil"))
prism_p30 <- rast(prism_paths[1])
prism_prec_2018 <- rast(prism_paths[2:13])
```

Packages like **tigris** and **prism** are designed to provide easy access to standard
datasets. Instead of downloading the data externally and then importing it
into R, these packages download the data within the R session. The **tigris**
package directly imports U.S. Census datasets into R objects, whereas the
prism package manages a library of downloaded raster datasets. There are
numerous other packages available that provide access to a variety of datasets.

Accessing remote data through packages can be very convenient. It also helps
to make scientific workflows more distributable and reproducible because
data acquisition and analysis can be integrated in a single R script. However,
ingtegrating remote data access into analytical workflows may not always be
the best approach. Downloading the data each time a script is run can be
time-consuming, and it may not be possible to run the script if there are
changes to the external data server, the server is inaccessible, or the internet
bandwidth is limited. In some cases, it may be more feasible to download the
data outside of the R session and then store and access them locally.

9.2   Zonal Statistics

Zonal statistics are a type of polygon-on-raster overlay in which the values in
the raster dataset are summarized within each polygon. The **terra** package
has a zonal() function for this type of calculation, but it requires a bit more
data preparation than when running zonal statistics in other GIS software
such as ArcGIS. This example will address the problem of summarizing mean
annual precipitation for every county in the United States.

To calculate zonal statistics, it is necessary to convert the vector polygon
dataset of counties to a raster dataset in which each raster cell is coded with
the 5-digit FIPS code of the corresponding county. The rasterize() function is

used to convert vector to raster data. The first argument is the county dataset
to summarize wrapped in the `vect()` function to convert it to the **terra** vector
data format. The second argument is a raster dataset to provide parameters
for the conversion. The output will match the CRS, extent, origin, numbers
of rows and columns, and cell size of the `prism_p30` dataset. This matching is
essential so that the rasterized counties and the PRISM precipitation grid can
be combined in the next step. The `field = "fips"` argument indicates which
column in the `county` vector dataset will be used to assign values to the raster
cells in the output.

```
cnty_ras <- rasterize(vect(county),
                      prism_p30,
                      field = "fips")
summary(cnty_ras)
##         fips
##  Min.    : 1001
##  1st Qu.:18177
##  Median :31091
##  Mean    :30738
##  3rd Qu.:46019
##  Max.    :56045
##  NA's    :45857
```

Zonal statistics are generated wtih the `zonal()` function. The first argument
specifies the raster layer to summarize and the second argument specifies the
zones to use for summarization. The `fun = "mean"` argument indicates the
summary function to use for each zone, and the `na.rm = TRUE` argument is
passed to the `mean()` function to remove `NA` values prior to summarization.
The output is a data frame with one row for each county. The precipitation
summary column initially has the long file name of the original image dataset
but is renamed as `precip` for simplicity.

```
cnty_p30 <- zonal(prism_p30,
                  cnty_ras,
                  fun = "mean",
                  na.rm = TRUE)
summary(cnty_p30)
##         fips          PRISM_ppt_30yr_normal_4kmM3_annual_bil
##  Min.    : 1001    Min.    :   78.88
##  1st Qu.:19041    1st Qu.: 763.73
##  Median :29205    Median :1095.83
##  Mean    :30625    Mean    :1026.01
##  3rd Qu.:45087    3rd Qu.:1277.06
##  Max.    :56045    Max.    :3068.14
```

```
cnty_p30 <- rename(cnty_p30,
                     precip = 2)
```

When processing data, it is always advisable to run basic checks to ensure that the results make sense. In this case, it is expected that the number of counties (rows) in the cnty_p30 datasets containing zonal summaries should match the number of counties in the original county dataset. However, this is not the case. There are fewer counties in the summary data table than there are counties in the polygon file.

```
dim(cnty_p30)
## [1] 3101     2
dim(county)
## [1] 3108    15
setdiff(county$fips,
        cnty_p30$fips)
## [1] 51580 51600 51678 51685 51683 51570 51610
```

9.3   Zone Size and Raster Cell Size

One of the most effective ways to assess data processing errors is through visualization. In this case, mapping the zonal summaries may reveal where data are missing. However, when the zonal summaries are joined to the county polygons and mapped, there are no obviously visible counties with missing data (Figure 9.1).

```
cnty_join1 <- left_join(county,
                        cnty_p30,
                        by = "fips")

ggplot(data = cnty_join1) +
  geom_sf(aes(fill = precip), size = 0.1) +
  scale_fill_continuous(name = "Precip (mm)") +
  theme_bw() +
  theme(legend.position = "bottom")
```

To make the map more readable, the scale_fill_distiller() function can be used to specify a yellow-to-green-to-blue color palette, change the direction so that areas with more precipitation are blue, and use a logarithmic scale instead of a linear scale for precipitation (Figure 9.2).

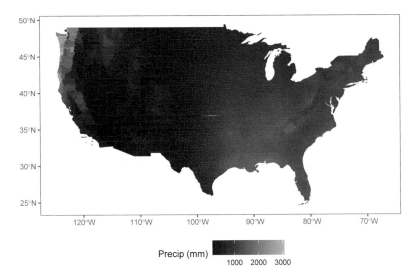

FIGURE 9.1
County-level zonal summaries of precipitation data from the PRISM dataset.

```
ggplot(data = cnty_join1) +
  geom_sf(aes(fill = precip), size = 0.1) +
  scale_fill_distiller(name = "Precip (mm)",
                       palette = "YlGnBu",
                       direction = 1,
                       trans = "log",
                       breaks = c(100, 300, 1000, 3000)) +
  theme_bw() +
  theme(legend.position = "bottom")
```

Displaying precipitation on a logarithmic scale is appropriate in this case because the data are heavily skewed, with most counties having relatively low values and a few having extremely large values. The logarithmic transformation stretches the lower values over a broader range of colors, which make the variation easier to see. However, the logarithm of zero is undefined and will return an NA value in R, so this approach does not work when there are zero values in the data.

Another way to look for the missing counties is to use the setdiff() function to return a vector of county FIPS codes that are present in the county dataset but not the cnty_p30 dataset. Note that all the missing counties have a state FIPS code of 51, which is Virginia.

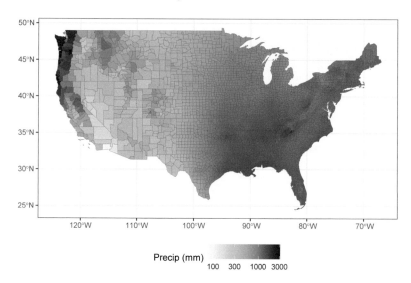

FIGURE 9.2
County-level zonal summaries of precipitation data from the PRISM dataset
with a logarithmic transformation.

```
setdiff(county$fips, cnty_p30$fips)
## [1] 51580 51600 51678 51685 51683 51570 51610
```

The next step is to zoom in and take a closer look at just Virginia. The
filter() function selects only the counties in Virginia, and a new map is
generated (Figure 9.3).

```
va_join1 <- filter(cnty_join1,
                   STATEFP == "51")

ggplot(data = va_join1) +
  geom_sf(aes(fill = precip), size = 0.1) +
  scale_fill_distiller(name = "Precip (mm)",
                       palette = "YlGnBu",
                       direction = 1,
                       trans = "log",
                       breaks = c(1000, 1150, 1300)) +
  theme_bw() +
  theme(legend.position = "bottom")
```

Virginia is unique among states in that it has many small, independent cities
that are governed separately from the surrounding counties and are therefore

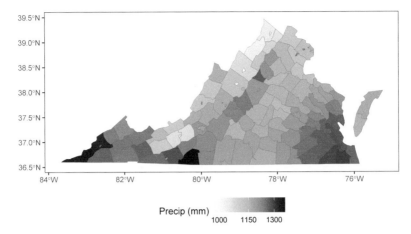

FIGURE 9.3
County-level zonal summaries of precipitation data from the PRISM dataset
for Virginia.

assigned their own county FIPS codes. However, it's still not clear where the
missing data are located. To zoom in further to a portion of Northern Virginia,
the xlim and ylim arguments can be specified in the coord_sf() function
(Figure 9.4).

```
ggplot(data = va_join1) +
  geom_sf(aes(fill = precip), size = 0.25) +
  coord_sf(xlim = c(-77.6, -77.0),
           ylim = c(38.6, 39.1)) +
  scale_fill_distiller(name = "Precip (mm)",
                       palette = "YlGnBu",
                       direction = 1,
                       trans = "log",
                       breaks = c(1000, 1150, 1300)) +
  theme_bw()
```

Now we can see several small cities with missing precipitation data. Why did
this happen? When the counties were rasterized to a 4 km grid to match the
precipitation data, the center point of each 4 km grid cell was sampled and
assigned the FIPS code of the county that it fell within. However, if a county
is very small or narrow, it is possible that no grid center point will fall within
its boundaries. In this situation, the county ends up excluded from the zonal
summary.

The simplest way to deal with this problem is to use a finer sampling grid
to generate the zonal statistics. This can be accomplished with the following

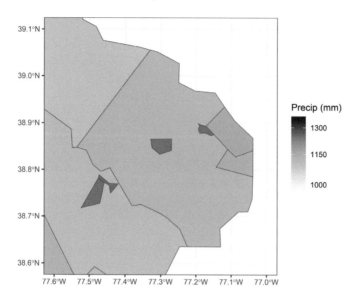

FIGURE 9.4

County-level zonal summaries of precipitation data from the PRISM dataset for northern Virginia.

steps. The disagg() function is used to reduce the cell size of the PRISM data by a factor of four, changing the cell size to approximately 1 km. Bilinear resampling is used to conduct linear interpolation between the center points of the 4 km grid cells to estimate values at the 1 km resolution. Then, the new 1 km grid is used to rasterize the county dataset, and the zonal() function is run using the 1 km county and precipitation rasters.

```
prism_p30_1km <- disagg(prism_p30,
                        fact = 4,
                        method = "bilinear")
cnty_ras_1km <- rasterize(vect(county),
                        prism_p30_1km,
                        field = "fips")
cnty_p30_1km <- zonal(prism_p30_1km,
                        cnty_ras_1km,
                        fun = "mean",
                        na.rm=T)
summary(cnty_p30_1km)
##      fips        PRISM_ppt_30yr_normal_4kmM3_annual_bil
##  Min.    : 1001   Min.    :   78.39
##  1st Qu.:19045    1st Qu.: 765.71
```

```
##  Median :29212    Median :1096.10
##  Mean    :30672   Mean     :1026.02
##  3rd Qu.:46008    3rd Qu.:1276.20
##  Max.    :56045   Max.     :3041.39
cnty_p30_1km <- rename(cnty_p30_1km,
                    precip = 2)
```

There is also a `resample()` function that does essentially the same thing but allows the user to specify a different grid for the resampling rather than calculating resolution as a factor of the current grid cell size. The `resample()` function is used in situations where one raster needs to be adjusted to match another raster with a different grid cell size or grid origin.

Now the zonal summary table contains the same number of records as the county file.

```
dim(cnty_p30_1km)
## [1] 3108    2
dim(county)
## [1] 3108   15
setdiff(county$fips,
        cnty_p30_1km$fips)
## numeric(0)
```

By checking the zoomed-in map of Virginia, it is apparent that all of the counties, including the tiny independent cities, now have precipitation values (Figure 9.5).

```
cnty_join2 <- left_join(county,
                      cnty_p30_1km,
                      by = "fips")
va_join2 <- filter(cnty_join2, STATEFP == "51")

ggplot(data = va_join2) +
  geom_sf(aes(fill = precip), size = 0.25) +
  scale_fill_distiller(name = "Precip (mm)",
                      palette = "YlGnBu",
                      direction = 1,
                      trans = "log",
                      breaks = c(1000, 1150, 1300)) +
  coord_sf(xlim = c(-77.6, -77.0),
            ylim = c(38.6, 39.1)) +
  theme_bw()
```

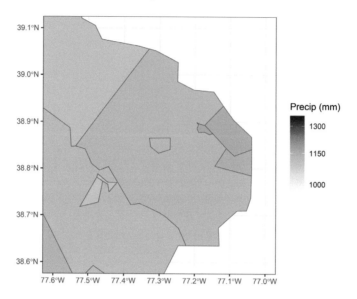

FIGURE 9.5
County-level zonal summaries of precipitation data from the PRISM dataset
for northern Virginia using raster with 1 km cell size.

It should be noted that the PRISM climate data can also be downloaded at
their native resolution of 800 m, and using these finer-resolution data would
be the best approach in real applications. The coarser 4 km dataset was used
in this example to reduce the downloaded data volume and to illustrate the
potential problems that can arise when there is a scale mismatch between two
datasets.

There are other scale issues that should be taken into account when working
with zonal statistics. As shown in the previous example, the sizes of U.S.
countries vary tremendously. Independent cities in Virginia may encompass
only a few square kilometers and thus have a relatively homogeneous climate.
In the western U.S., counties are much larger and more heterogeneous. For
example, King County Washington covers thousands of square kilometers
ranging from the coast to the crest of the Cascade Mountain Range. Most
people live in the coastal regions in and around the city of Seattle, and zonal
summaries across the entire county are not a precise metric of the climates that
they experience. This is one aspect of the well-known Modifiable Areal Unit
Problem (MAUP) in geography (Dark and Bram, 2007). Before conducting
a zonal analysis, it is important to carefully consider the sizes and spatial
patterns of the zones and ensure that they are appropriate for the specific
question at hand.

9.4 Extracting Raster Values with Point Data

In other situations, it is necessary to extract raster cell values associated with specific point locations. The following example will assess the accuracy of the monthly PRISM precipitation datasets for 2018 by comparing them with weather station data. These data were previously downloaded using the **prism** package and imported into the `prism_prec_2018` dataset.

The PRISM data will be compared to monthly station data from the small Oklahoma mesonet data file, `mesodata_small.csv`, which was introduced in Chapter 2. The geographic coordinates of these stations also need to be imported separately from the `geoinfo.csv` file.

```
mesosm <- read_csv("mesodata_small.csv")
geo_coords <- read_csv("geoinfo.csv")
```

The data frame containing the geographic coordinates is converted to an `sf` object using `st_as_sf()`. The `unique()` function returns a vector of the station ID codes in the Mesonet dataset, and a new `sf` object containing only the point data for these stations is generated using the `filter()` function.

```
geo_coords <- st_as_sf(geo_coords,
                       coords = c("lon", "lat"))
mystations <- unique(mesosm$STID)
station_pts <- geo_coords %>%
  filter(stid %in% mystations)
```

The `extract()` function is used to obtain the raster data associated with each of the four points. First, the multilayer raster containing the precipitation data is assigned layer names. Since there is one layer for each month, we can use `month.abb`, a built-in R constant that contains vector of twelve abbreviations for the months of the year.

```
month.abb
##  [1] "Jan" "Feb" "Mar" "Apr" "May" "Jun" "Jul" "Aug" "Sep"
## [10] "Oct" "Nov" "Dec"
names(prism_prec_2018) <- month.abb
prism_samp <- extract(prism_prec_2018,
                      vect(station_pts),
                      factors = T,
                      df = T)
```

There are twelve layers in the `prism_prec_2018` raster stack, one for each month. Therefore, the table generated by `extract()` has 48 values (4 stations x 12 months).

```
dim(prism_samp)
## [1]   4 13
glimpse(prism_samp)
## Rows: 4
## Columns: 13
## $ ID  <dbl> 1, 2, 3, 4
## $ Jan <dbl> 0.000, 82.706, 10.658, 7.660
## $ Feb <dbl> 1.464, 343.607, 104.995, 78.992
## $ Mar <dbl> 0.000, 128.251, 31.580, 13.819
## $ Apr <dbl> 32.591, 128.465, 38.255, 64.245
## $ May <dbl> 53.662, 140.197, 96.068, 124.110
## $ Jun <dbl> 129.731, 75.741, 81.546, 204.971
## $ Jul <dbl> 93.196, 93.599, 70.848, 88.683
## $ Aug <dbl> 46.675, 160.207, 56.496, 69.179
## $ Sep <dbl> 37.529, 140.801, 59.430, 153.339
## $ Oct <dbl> 98.560, 163.322, 101.636, 107.868
## $ Nov <dbl> 8.252, 112.287, 30.242, 13.804
## $ Dec <dbl> 57.315, 174.494, 86.176, 112.201
```

The `prism_samp` data frame is in wide format, with one row for each station and one column for each month of precipitation data plus an additional column that contains a numerical ID code. The `mesosm` data frame is in a long format, with one row for each combination of station, month, and year. Before comparing the month precipitation values from the two data sources, they must be reformatted and combined into a single data frame. There are a number of steps in this process, but they can be coded concisely using a series of piped **dplyr** and **tidyr** functions as described in Chapter 3.

The `bind_cols()` function is used to add the columns of the original station location data frame, `station_pts` to `prism_samp`. This function only works if the two data frames have the same number of rows, and the columns of the second data frame are simply appended to the first. This step is necessary to add the station ID codes, which will be needed later.

The `pivot_longer` function is used to combine all twelve precipitation columns, `Jan` through `Dec`, into a new column called `PPrec` while the corresponding month abbreviations are stored in `mnth_name`.

The `mutate()` function adds two new columns: `month` contains numeric codes for each month that are generated with the `match()` function, and `PPrec_in` contains the PRISM precipitation values converted to inches.

The `inner_join()` function is used to join these data with mesosm, which contains the Mesonet data from the stations by station ID code and month number.

Finally, the `filter()` function extracts only the 2018 records, and the `select()` function retains only the columns that are needed for subsequent graphing and analysis.

```
compare_prec <- prism_samp %>%
  bind_cols(station_pts) %>%
  pivot_longer(Jan:Dec,
               names_to = "mnth_name",
               values_to = "PPrec") %>%
  mutate(month = match(mnth_name, month.abb),
         PPrec_in = PPrec * 0.0393701) %>%
  inner_join(mesosm, by = c("stid" = "STID", "month" = "MONTH")) %>%
  filter(YEAR == 2018) %>%
  select(stid, month, RAIN, PPrec_in)
```

The critical variables in the resulting data frame are the monthly rainfall measured at the mesonet stations (`RAIN`) and the monthly rainfall estimates from the PRISM dataset converted to inches to match the Mesonet data (`Pprec_in`) along with the station ID (`stid`) and month (`month`) codes.

```
glimpse(compare_prec)
## Rows: 48
## Columns: 4
## $ stid     <chr> "HOOK", "HOOK", "HOOK", "HOOK", "HOOK", "~
## $ month    <dbl> 1, 2, 3, 4, 5, 6, 7, 8, 9, 10, 11, 12, 1,~
## $ RAIN     <dbl> 0.00, 0.06, 0.00, 1.25, 1.92, 5.07, 2.85,~
## $ PPrec_in <dbl> 0.00000000, 0.05763783, 0.00000000, 1.283~
```

The `compare_prec` data frame can now be used to generate a scatterplot showing the relationship between Mesonet weather station data on the x-axis and gridded PRISM data on the y-axis (Figure 9.6). The `geom_abline()` function is used to add a dashed 1:1 line to the graph. Of the four stations, Mt. Herman (MTHE) appears to have the largest deviations from the 1:1 line.

```
ggplot(data = compare_prec) +
  geom_point(aes(x = RAIN,
                 y = PPrec_in,
                 color = stid)) +
  scale_color_discrete(name = "Station ID") +
  geom_abline(slope = 1,
```

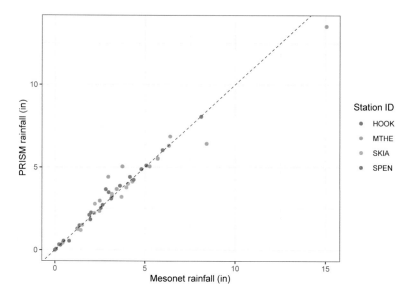

FIGURE 9.6
Scatterplot of monthly 2018 precipitation data from the Oklahoma Mesonet versus PRISM for four stations in Oklahoma.

```
                 intercept = 0,
                 size = 0.25,
                 linetype = "dashed") +
  xlab("Mesonet rainfall (in)") +
  ylab("PRISM rainfall (in)") +
  theme_bw()
```

The following code generates some additional summary statistics for the relationships between these precipitation estimates. The linear regression has a high R-squared and a slope that is only slightly less than one.

```
rain_lm <- lm(PPrec_in ~ RAIN, data = compare_prec)
summary(rain_lm)
##
## Call:
## lm(formula = PPrec_in ~ RAIN, data = compare_prec)
##
## Residuals:
##      Min       1Q    Median       3Q      Max
## -1.56891 -0.25019 -0.06633  0.20765  1.37211
```

```
##
## Coefficients:
##              Estimate Std. Error t value Pr(>|t|)
## (Intercept)   0.31273    0.10799   2.896  0.00577 **
## RAIN          0.91502    0.02528  36.196  < 2e-16 ***
## ---
## Signif. codes:
## 0 '***' 0.001 '**' 0.01 '*' 0.05 '.' 0.1 ' ' 1
##
## Residual standard error: 0.466 on 46 degrees of freedom
## Multiple R-squared:  0.9661, Adjusted R-squared:  0.9653
## F-statistic:  1310 on 1 and 46 DF,  p-value: < 2.2e-16
```

Other accuracy statistics such as the root mean squared error (RMSE), mean absolute error (MAE), and mean error (ME) are calculated using the sum-marize() function. These all have relatively low values, indicating a close agreement between the PRISM and Mesonet rainfall measurements.

```
rain_sum <- compare_prec %>%
  summarize(RMSE = sqrt(mean((PPrec_in - RAIN)^2)),
            MAE = mean(abs(PPrec_in - RAIN)),
            ME = mean(PPrec_in - RAIN))
rain_sum
## # A tibble: 1 x 3
##    RMSE   MAE     ME
##   <dbl> <dbl>  <dbl>
## 1 0.510 0.283 0.0287
```

There seems to be a very strong relationship between the two data sources. However, the Oklahoma Mesonet dataset is one of the sources of station data that are used to develop the interpolated PRISM dataset. So the two datasets are not entirely independent. Although the PRISM precipitation estimates are well calibrated to the Mesonet observations, this relationship may not be representative of PRISM accuracy at other locations where there are no Mesonet stations.

9.5 Practice

1. Starting with the sf object containing the county-level precipitation summaries, convert the precipitation units from millimeters to inches. Then classify precipitation into a factor with five

categories: 0-10, 10-35, 35-50, 50-70, and > 70 inches. Generate a new map of precipitation using these categories.

2. Generate a map of mean annual temperature summarized by county for the conterminous United States. You can access the PRISM mean temperature data by using `get_prism_normals()` and specifying `type = 'tmean'`. Then you will need to recreate the `prism_paths` vector and select the appropriate element to import the temperature raster. Compute zonal summaries of mean temperature for every county and create a map to display the results. The distribution of temperature is much less skewed than that of precipitation, so a logarithmic transformation is not needed.

3. Repeat the comparison of gridded PRISM data and Mesonet station data using maximum temperature. You can access the PRISM maximum temperature data by using the `get_prism_monthly()` and specifying `type = 'tmax'`. The corresponding variable in the Mesonet dataset is `TMAX`.

10

Combining Vector Data with Discrete Raster Data

This chapter will introduce several new datasets and additional examples of integrating vector and raster data. The Cropland Data Layer (CDL) is a categorical raster dataset that maps agricultural land use in the United States. It provides annual data on locations where specific crop types such as corn, soybeans, wheat, and many others are planted. These maps are produced by the United States Department of Agriculture (USDA), which collects massive amounts of field-level data on planted crops from farmers. The data are used to train machine learning algorithms that predict the spatial pattern of crop types using satellite remote sensing data (Boryan et al., 2011). The CDL datasets can be visualized and accessed through the CropScape online platform (`https://nassgeodata.gmu.edu/CropScape/`).

The National Hydrography Dataset (NHD) contains vector data on the water drainage network in the United States. It consists of points, lines, and polygons that represent hydrological unit boundaries as well as features such as lakes, ponds, rivers, and streams. These data can be downloaded through the United State Geological Survey (USGS) National Map (`https://apps.nationalmap.gov/downloader/`).

This chapter will use downloaded data files characterizing crop types, hydrological unit boundaries, and stream data from the Middle Big Sioux Subbasin in eastern South Dakota. However, these data can also be accessed directly through R using the **CropScapeR** (Chen, 2021) and **nhdR** (Stachelek, 2022) packages.

Analyzing changes in the relative abundance of crop types and other land cover classes is important for understanding how agricultural land use is changing over time. In eastern South Dakota and other locations at the edge of the U.S. "corn belt," there is concern that expanding cropland is reducing the cover of grasslands, including lands managed for pasture and hay as well as native prairie remnants (Wimberly et al., 2017). Knowing whether these changes are occurring close to streams and other water bodies is important for assessing their impacts on wildlife habitat, carbon storage, and water quality.

The required R packages have all been used before with the exception of **ggspatial** (Dunnington, 2022), which provides some additional functions for

DOI: 10.1201/9781003326199-10

map annotations when using `ggplot()`. The `rasterdf()` function is also sourced for mapping the raster data in `ggplot()`.

```
library(tidyverse)
library(sf)
library(terra)
library(scales)
library(colorspace)
library(ggspatial)
source("rasterdf.R")
```

The raster data include TIF files for 2010 and 2020 CDL data covering a portion of eastern South Dakota that were downloaded using the CropScape website. The vector data include shapefiles with flowlines, watershed boundaries, and subwatershed boundaries within the Middle Big Sioux subbasin in eastern South Dakota.

```
# Read in the data
crop2010 <- rast("CDL_2010_clip.tif")
crop2020 <- rast("CDL_2020_clip.tif")
streams <- st_read("NHDFlowline.shp", quiet = TRUE)
wsheds <- st_read("WBDHU10.shp", quiet = TRUE)
subwsh <- st_read("WBDHU12.shp", quiet = TRUE)
```

Before trying to combine the two data sources, it is important to check their coordinate reference systems. The NHD data are in a geographic coordinate system.

```
writeLines(st_crs(wsheds)$WktPretty)
## GEOGCS["NAD83",
##     DATUM["North_American_Datum_1983",
##         SPHEROID["GRS 1980",6378137,298.257222101]],
##     PRIMEM["Greenwich",0],
##     UNIT["degree",0.0174532925199433,
##         AUTHORITY["EPSG","9122"]],
##     AXIS["Latitude",NORTH],
##     AXIS["Longitude",EAST],
##     AUTHORITY["EPSG","4269"]]
```

The CDL data are in a projected Albers Equal Area coordinate system.

```
writeLines(st_crs(crop2010)$WktPretty)
## PROJCS["unnamed",
```

```
##      GEOGCS["NAD83",
##          DATUM["North_American_Datum_1983",
##              SPHEROID["GRS 1980",6378137,298.257222101004]],
##          PRIMEM["Greenwich",0],
##          UNIT["degree",0.0174532925199433,
##              AUTHORITY["EPSG","9122"]],
##          AUTHORITY["EPSG","4269"]],
##      PROJECTION["Albers_Conic_Equal_Area"],
##      PARAMETER["latitude_of_center",23],
##      PARAMETER["longitude_of_center",-96],
##      PARAMETER["standard_parallel_1",29.5],
##      PARAMETER["standard_parallel_2",45.5],
##      PARAMETER["false_easting",0],
##      PARAMETER["false_northing",0],
##      UNIT["metre",1,
##          AUTHORITY["EPSG","9001"]],
##      AXIS["Easting",EAST],
##      AXIS["Northing",NORTH]]
```

For the reasons discussed in Chapter 8, all three vector datasets are reprojected to match the coordinate system of the raster data rather than reprojecting the raster data.

```
wsheds_aea <- st_transform(wsheds, crs(crop2010))
subwsh_aea <- st_transform(subwsh, crs(crop2010))
streams_aea <- st_transform(streams, crs(crop2010))
```

10.1   Visualizing and Manipulating Vector Data

The first step is to explore the watersheds within the Middle Big Sioux Subbasin by generating a labeled map. First, columns with x and y coordinates are added to the wsheds object to make it possible to label the watersheds. For map labels, it is advisable to use st_point_on_surface() instead of st_centroid() to ensure that the label points fall inside the polygons. The st_coordinates() function converts these coordinates into a geometry column that can be interpreted by **sf** functions. Then a labeled map of watersheds is generated by using geom_sf() to plot the watershed boundaries (the default fill color is gray) and using geom_text() to write the name of each watershed at its label point (Figure 10.1).

FIGURE 10.1
Watersheds in the Middle Big Sioux subbasin.

```
wsheds_aea <- cbind(wsheds_aea,
                st_coordinates(st_point_on_surface(wsheds_aea)))

ggplot(data = wsheds_aea) +
  geom_sf(color = "black",
          size = 0.1) +
  geom_text(aes(x = X,
                y = Y,
                label = name),
            size = 3,
            fontface = "bold") +
  coord_sf() +
  theme_void()
```

Some of the labels are crowded and difficult to read. Also, because of the
complex shapes of some of the watersheds, the point locations generated by
st_point_on_surface() are not well centered. When there are relatively few

labels as in this example, it is often possible to adjust the text placement manually. This is the order of watersheds in the wsheds data frame.

```
wsheds$name
##  [1] "Lake Sinai-Big Sioux River"
##  [2] "Deer Creek"
##  [3] "Medary Creek"
##  [4] "Dry Lake Number One"
##  [5] "Lake Marsh"
##  [6] "Lake Poinsett"
##  [7] "Hidewood Creek"
##  [8] "Oakwood Lakes"
##  [9] "Sixmile Creek"
## [10] "North Deer Creek"
## [11] "Battle Creek"
```

The following code adjusts the text positions for "Lake Sinai-Big Sioux River" (position 1), "Deer Creek (position 2), and North Deer Creek (position 10). The hnudge and vnudge vectors contain the horizontal and vertical adjustments for all eleven watersheds. These values are in map units, which in this case are meters. Zero values are provided when the positions will not be adjusted. Then, the nudge_x = hnudge and nudge_y = vnudge arguments to geom_text() are specified when generating the map text. Further refinements could be made, but the text here is at least a bit more readable (Figure 10.2).

```
hnudge <- c(-10000, 5000, 0, 0, 0, 0, 0, 0, 0, 0, 0)
vnudge <- c(-7500, 2500, 0, 0, 0, 0, 0, 0, 0, 5000, 0)

ggplot(data = wsheds_aea) +
  geom_sf(color = "black",
          size = 0.1) +
  geom_text(aes(x = X,
                y = Y,
                label = name),
            nudge_x = hnudge,
            nudge_y = vnudge,
            size = 3,
            fontface = "bold") +
  theme_void()
```

The label positions can also be modified using an alternative geom function. The geom_spatial_label_repel() function from the **ggspatial** package automatically modifies the label placement to keep them from overlapping (Figure 10.3). The argument label = str_wrap(name, 15) forces the labels to

FIGURE 10.2
Watersheds in the Middle Big Sioux subbasin with manually adjusted label positions.

wrap so that the maximum width is 15 characters, which makes them more compact.

```
ggplot(data = wsheds_aea) +
  geom_sf(color = "black",
          size = 0.1) +
  geom_spatial_label_repel(aes(x = X,
                               y = Y,
                               label = str_wrap(name, 15)),
                           size = 3,
                           fontface = "bold",
                           crs = st_crs(wsheds_aea)) +
  theme_void()
```

The next map includes the watershed boundaries as well as the nested sub-watershed boundaries displayed in dark gray (Figure 10.4). Because the data are from multiple data frames, the data argument needs to be included in the geom_sf() functions instead of the initial ggplot() function.

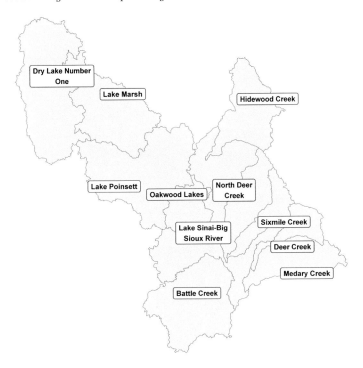

FIGURE 10.3
Watersheds in the Middle Big Sioux subbasin with automatically adjusted label positions.

The `annotation_scale()` function from the **ggspatial** package is also used to add a scale bar to better discern the sizes of the polygons.

```
ggplot() +
  geom_sf(data = subwsh_aea, color = "darkgray") +
  geom_sf(data = wsheds_aea, color = "black", fill = NA) +
  annotation_scale(location = 'bl') +
  theme_void()
```

The following analyses will focus on the North Deer Creek watershed. To extract the boundary for only this watershed, the `filter()` function is used.

```
wshed_ndc <- wsheds_aea %>%
  filter(name == "North Deer Creek")
```

FIGURE 10.4
Watersheds and subwatersheds in the Middle Big Sioux subbasin.

A second `filter()` function with a nested `sf_covered_by()` function is then used to make a spatial query that selects all subwatershed polygons in the `subwsh_aea` layer that fall within the North Deer Creek watershed.

```
subwsh_ndc <- subwsh_aea %>%
  filter(lengths(st_covered_by(., wshed_ndc)) > 0)
```

10.2   Zonal Summaries of Discrete Raster Data

To summarize the crop type data for each subwatershed, the CDL data are first cropped and masked to the boundaries of the North Deer Creek watershed.

```
cdlstack <- c(crop2010, crop2020)
names(cdlstack) <- c("2010", "2020")

cdl_crop <- crop(cdlstack, vect(subwsh_ndc))
subwsh_msk <- rasterize(vect(subwsh_ndc), cdl_crop)
cdl_mbs <- mask(cdl_crop, subwsh_msk)
```

The approach that was introduced in Chapter 7 is used to reclassify the CDL into a smaller number of classes. The most widespread crops in this region are corn (1) and soybeans (5). Grasslands and pastures (37) non-alfalfa hay (176) and herbaceous wetlands (195) are combined into a single grass class. The `others = 4` argument in the `classify()` function is used to combine all the other codes into a single class.

```
oldclas <- c(1, 5, 37, 176, 195)
newclas <- c(1, 2, 3, 3, 3)
lookup <- data.frame(oldclas, newclas)
cdl_rc <- classify(cdl_mbs,
                       rcl = lookup,
                       others = 4)
```

The reclassified rasters are then plotted with the subwatershed boundaries overlaid (Figure 10.5).

```
newnames <- c("Corn",
                  "Soybeans",
                  "Grass",
                  "Other")
newcols <- c("yellow",
                "green",
                "tan",
                "gray60")
newcols2 <- desaturate(newcols,
                         amount = 0.2)
cdl_df <- rasterdf(cdl_rc)

ggplot(data = cdl_df) +
  geom_raster(aes(x = x,
                    y = y,
                    fill =
                        as.character(value))) +
  scale_fill_manual(name = "Crop Type",
                        values = newcols2,
```

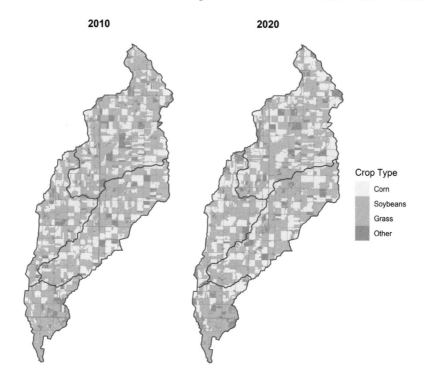

FIGURE 10.5
Subwatershed boundaries overlaid on the cropland data layer in the North
Deer Creek watershed

```
                labels = newnames,
                na.translate = FALSE) +
geom_sf(data = subwsh_ndc,
        fill = NA) +
facet_wrap(facets = vars(variable), ncol = 2) +
theme_void() +
theme(strip.text.x = element_text(size=12, face="bold"))
```

The zonal() function, introduced in Chapter 9, is useful for generating statistical summaries of continuous data. However, a different approach is needed to summarize discrete data, where we want to estimate the areas of different classes within each polygon. We start by rasterizing the subwatershed polygons to match the raster grid of the land cover data. Then the crosstab() function, which was used for change analysis in Chapter 7, is used to calculate the number of cells of each land cover type within each subwatershed polygon.

```
subwsh_r <- rasterize(vect(subwsh_ndc),
                      cdl_rc,
                      field = "ObjectID")
cdlsum10 <- crosstab(c(subwsh_r, cdl_rc[["2010"]]))
cdlsum10 <- as_tibble(cdlsum10)
cdlsum20 <- crosstab(c(subwsh_r, cdl_rc[["2020"]]))
cdlsum20 <- as_tibble(cdlsum20)
```

Next, a series of `dplyr()` and `tidyr()` commands are used to convert the crosstabulation results into a long format for mapping and computing the percent area of each crop type in each subwatershed. The process will be broken down into multiple steps.

The `inner_join()` function is used to join the 2010 and 2020 crosstabulation results by the subwatershed code (stored in `ObjectID`) and the crop type code (stored in `X2010` and `X2020`).

```
cdljoin <- cdlsum10 %>%
  inner_join(cdlsum20,
             by = c("ObjectID" = "ObjectID",
                    "X2010" = "X2020"))
cdljoin
## # A tibble: 16 x 4
##      ObjectID X2010   n.x   n.y
##      <chr>    <chr> <int> <int>
##  1 38        1     42810 53551
##  2 39        1     22745 26070
##  3 40        1     35535 37192
##  4 41        1      6603  8636
##  5 38        2     48192 43019
##  6 39        2     22238 19146
##  7 40        2     40287 41062
##  8 41        2      8704  6928
##  9 38        3     32396 25490
## 10 39        3     13584 12382
## 11 40        3     19865 16614
## 12 41        3     17892 16703
## 13 38        4     21846 23184
## 14 39        4      7172  8141
## 15 40        4     13298 14117
## 16 41        4      7155  8087
```

The resulting table has the crop type codes stored in the `X2010` column. The `mutate()` function is used to convert these to a new `croptype` column that is a

factor with labels for each of the crop type names. The old X2010 column is
discarded. New names are assigned to the n.x and n.y columns, which contain
the counts of the 2010 and 2020 crop types for each subwatershed.

```
cdljoin2 <- cdljoin %>%
  mutate(croptype = factor(X2010,
                           labels = newnames),
         ObjectID = as.numeric(ObjectID)) %>%
  select(-X2010) %>%
  rename("cnt2010" = "n.x",
         "cnt2020" = "n.y")
cdljoin2
## # A tibble: 16 x 4
##     ObjectID cnt2010 cnt2020 croptype
##        <dbl>   <int>   <int> <fct>
## 1        38   42810   53551 Corn
## 2        39   22745   26070 Corn
## 3        40   35535   37192 Corn
## 4        41    6603    8636 Corn
## 5        38   48192   43019 Soybeans
## 6        39   22238   19146 Soybeans
## 7        40   40287   41062 Soybeans
## 8        41    8704    6928 Soybeans
## 9        38   32396   25490 Grass
## 10       39   13584   12382 Grass
## 11       40   19865   16614 Grass
## 12       41   17892   16703 Grass
## 13       38   21846   23184 Other
## 14       39    7172    8141 Other
## 15       40   13298   14117 Other
## 16       41    7155    8087 Other
```

The pivot_longer() function is used to combine the 2010 and 2020 cell counts
into a single cnt column with year information stored in the year column. The
parse_number() function is then used to convert year from a character to a
number.

```
cdlpivot <- cdljoin2 %>%
  pivot_longer(cols = starts_with("cnt"),
               names_to = "year",
               values_to = "cnt") %>%
  mutate(year = parse_number(year))
cdlpivot
## # A tibble: 32 x 4
```

```
##      ObjectID croptype  year    cnt
##         <dbl> <fct>    <dbl>  <int>
## 1         38 Corn       2010 42810
## 2         38 Corn       2020 53551
## 3         39 Corn       2010 22745
## 4         39 Corn       2020 26070
## 5         40 Corn       2010 35535
## 6         40 Corn       2020 37192
## 7         41 Corn       2010  6603
## 8         41 Corn       2020  8636
## 9         38 Soybeans   2010 48192
## 10        38 Soybeans   2020 43019
## # ... with 22 more rows
```

The total number of pixels for each combination of subwatershed and year is summed, and the percent area of each crop type within each subwatershed is calculated.

```
cdlperc <- cdlpivot %>%
  group_by(ObjectID, year) %>%
  mutate(tot = sum(cnt),
         perc_crop = 100 * cnt / tot)
cdlperc
## # A tibble: 32 x 6
## # Groups:   ObjectID, year [8]
##      ObjectID croptype  year    cnt      tot perc_crop
##         <dbl> <fct>    <dbl>  <int>    <int>     <dbl>
## 1         38 Corn       2010 42810  145244      29.5
## 2         38 Corn       2020 53551  145244      36.9
## 3         39 Corn       2010 22745   65739      34.6
## 4         39 Corn       2020 26070   65739      39.7
## 5         40 Corn       2010 35535  108985      32.6
## 6         40 Corn       2020 37192  108985      34.1
## 7         41 Corn       2010  6603   40354      16.4
## 8         41 Corn       2020  8636   40354      21.4
## 9         38 Soybeans   2010 48192  145244      33.2
## 10        38 Soybeans   2020 43019  145244      29.6
## # ... with 22 more rows
```

Note that the same code can also be implemented more concisely as a single block of piped functions.

```
cdlperc <- cdlsum10 %>%
  inner_join(cdlsum20,
             by = c("ObjectID" = "ObjectID",
                    "X2010" = "X2020")) %>%
  mutate(croptype = factor(X2010,
                           labels = newnames),
         ObjectID = as.numeric(ObjectID)) %>%
  rename("cnt2010" = "n.x",
         "cnt2020" = "n.y") %>%
  pivot_longer(cols = starts_with("cnt"),
               names_to = "year",
               values_to = "cnt") %>%
  mutate(year = parse_number(year)) %>%
  group_by(ObjectID, year) %>%
  mutate(tot = sum(cnt),
         perc_crop = 100 * cnt / tot)
```

The resulting table can be joined to the data frame containing the subwatershed polygons and used to map the percent of each crop type by subwatershed in 2010 and 2020 (Figure 10.6). The `facet_wrap()` function with the formula `year ~ croptype` is used to create a layout with years as rows and crop types as columns.

```
subwsh_perc <- left_join(subwsh_ndc,
                         cdlperc,
                         by = "ObjectID")

ggplot(data = subwsh_perc) +
  geom_sf(aes(fill = perc_crop)) +
  scale_fill_distiller(name="Percent",
                       palette = "YlGn",
                       breaks = pretty_breaks(),
                       direction = 1) +
  facet_wrap(year ~ croptype,
             ncol = 4) +
  theme_void() +
  theme(strip.text.x = element_text(size=12, face="bold"))
```

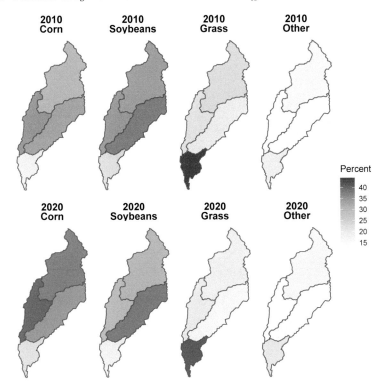

FIGURE 10.6
Proportional area of each crop type summarized by subwatershed in the North
Deer Creek watershed.

10.3 Summarizing Land Cover With Stream Buffers

The streamlines from the NHD are read in as three-dimensional data with Z
coordinates, and they need to be converted to two-dimensional data before
we can work with them. This step is accomplished using the st_zm() function.
The st_intersection() function is then used to clip the streams to the North
Deer Creek watershed boundary, and the result is plotted (Figure 10.7).

```
streams_aea <- st_zm(streams_aea)
streams_ndc <- st_intersection(streams_aea, wshed_ndc)

ggplot() +
  geom_sf(data = subwsh_ndc,
          color = "black",
```

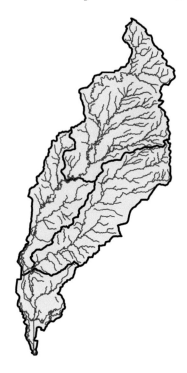

FIGURE 10.7
NHD streams in the North Deer Creek watershed.

```
        size = 1) +
  geom_sf(data = streams_ndc,
        color = "blue") +
  theme_void()
```

Stream buffers are computed with the st_buffer() function. The dist argument provides the buffer width in map units, which in this case are meters. The output contains a separate buffer polygon for each line segment in the stream network (Figure 10.8).

```
streams_buff <- st_buffer(streams_ndc, dist = 100)

ggplot() +
  geom_sf(data = wshed_ndc,
        color = "black",
        size = 1) +
  geom_sf(data = streams_buff,
```

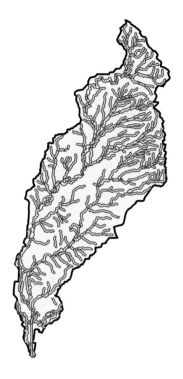

FIGURE 10.8
Buffers for each stream segment in the North Deer Creek watershed.

```
          color = "blue") +
  theme_void()
```

The st_union() function is used to combine the separate buffer polygons into a single polygon. Note that st_union() creates a simple feature geometry list column (sfc) that needs to be converted into an sf object before undertaking further analysis.

```
streams_comb <- st_union(streams_buff)
class(streams_comb)
## [1] "sfc_MULTIPOLYGON" "sfc"
streams_comb <- st_sf(streams_comb)
class(streams_comb)
## [1] "sf"           "data.frame"
```

Where streams are close to the watershed boundary, their buffers may extend outside of the watershed. The st_intersection() function combines the stream buffers with the North Deer Creek watershed polygon to create a polygon encompassing all stream buffers inside the watershed. The st_difference() function similarly creates a polygon encompassing the parts of the North Deer Creek watershed that are outside the stream buffers.

```
riparian <- st_intersection(streams_comb, wshed_ndc)
upland <- st_difference(wshed_ndc, streams_comb)
```

The stream buffers can be mapped as polygon boundaries overlaid on top of the reclassified land cover dataset (Figure 10.9).

```
ggplot(data = cdl_df) +
  geom_raster(aes(x = x,
                  y = y,
                  fill = as.character(value))) +
  scale_fill_manual(name = "Land cover",
                    values = newcols2,
                    labels = newnames,
                    na.translate = FALSE) +
  facet_wrap(facets = vars(variable), ncol = 2) +
  geom_sf(data = streams_comb,
          color = "darkblue",
          fill = NA) +
  coord_sf(expand = F) +
  theme_void() +
  theme(strip.text.x = element_text(size=12, face="bold"))
```

Two new raster datasets are created by masking the CDL dataset with the riparian and upland vector datasets.

```
riparian_mask <- rasterize(vect(riparian), cdl_rc)
cdl_rip <- mask(cdl_rc, riparian_mask)
upland_mask <- rasterize(vect(upland), cdl_rc)
cdl_up <- mask(cdl_rc, upland_mask)
```

The freq() function is used to compute the pixel counts in the resulting riparian and upland raster datasets.

```
freq_rip <- as_tibble(freq(cdl_rip))
freq_rip
## # A tibble: 8 x 3
##    layer value count
##    <dbl> <dbl> <dbl>
```

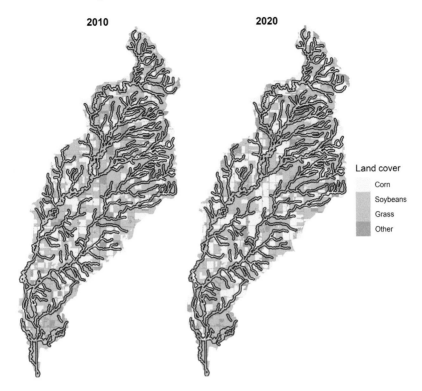

2010   **2020**

Land cover
- Corn
- Soybeans
- Grass
- Other

FIGURE 10.9
Stream buffers overlaid on crop types for the North Deer Creek watershed.

```
## 1       1      1 26599
## 2       1      2 26396
## 3       1      3 45230
## 4       1      4 12117
## 5       2      1 29395
## 6       2      2 24737
## 7       2      3 40868
## 8       2      4 15342
freq_up <- as_tibble(freq(cdl_up))
freq_up
## # A tibble: 8 x 3
##    layer value count
##    <dbl> <dbl> <dbl>
## 1      1      1 81094
## 2      1      2 93025
## 3      1      3 38507
```

```
## 4        1        4 37354
## 5        2        1 96054
## 6        2        2 85418
## 7        2        3 30321
## 8        2        4 38187
```

The outputs are combined using `bind_rows()` into a single data frame that
contains pixel counts for each land cover class for the buffer and upland
zones. This function will only work if the two data frames have the exact
same column specifications, and it "stacks" them on top of one another. The
`landtype` column is created to identify the riparian and upland records, and
the `year` and `croptype` columns are derived from the `layer` and `value` columns
output by the `freq()` function.

The pixel counts are used to calculate the percent cover for each subwatershed.
First, the data are grouped by `landtype` (stream buffer versus upland), and
then the `mutate()` function is run on the grouped data. Note that the results
are different from the `summarize()` function. The `mutate()` function replicates
the results of a summary (in this case, the `sum()` function) for each record
within a group rather than returning a single value for each group. Having the
grouped total of pixel counts on each line makes it straightforward to calculate
the percent cover of each class.

```
cdl_chng <- freq_rip %>%
  bind_rows(freq_up) %>%
  mutate(landtype = c(rep("Riparian", 8), rep("Upland", 8)),
         year = factor(layer,
                       labels = c("2010", "2020")),
         croptype = factor(value,
                           labels = newnames)) %>%
  select(-layer, -value) %>%
  group_by(landtype) %>%
  mutate(totarea = sum(count),
         percarea = 100 * count / totarea)
cdl_chng
## # A tibble: 16 x 6
## # Groups:   landtype [2]
##    count landtype year  croptype totarea percarea
##    <dbl> <chr>    <fct> <fct>      <dbl>    <dbl>
## 1 26599 Riparian 2010  Corn      220684     12.1
## 2 26396 Riparian 2010  Soybeans  220684     12.0
## 3 45230 Riparian 2010  Grass     220684     20.5
## 4 12117 Riparian 2010  Other     220684      5.49
## 5 29395 Riparian 2020  Corn      220684     13.3
```

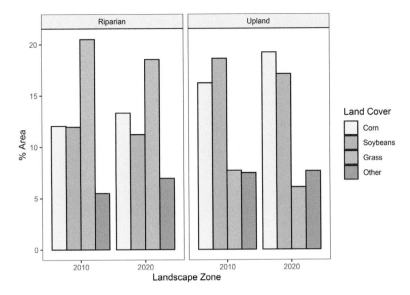

FIGURE 10.10
Areal distribution of crop types by landscape zone (riparian versus upland)
and year.

```
##  6 24737 Riparian 2020  Soybeans  220684  11.2
##  7 40868 Riparian 2020  Grass     220684  18.5
##  8 15342 Riparian 2020  Other     220684   6.95
##  9 81094 Upland   2010  Corn      499960  16.2
## 10 93025 Upland   2010  Soybeans  499960  18.6
## 11 38507 Upland   2010  Grass     499960   7.70
## 12 37354 Upland   2010  Other     499960   7.47
## 13 96054 Upland   2020  Corn      499960  19.2
## 14 85418 Upland   2020  Soybeans  499960  17.1
## 15 30321 Upland   2020  Grass     499960   6.06
## 16 38187 Upland   2020  Other     499960   7.64
```

These data can be graphed as a bar chart to compare the distribution of
crop types by year and by riparian or upland landscape zone (Figure 10.10).
Riparian areas have more grass cover relative to corn and soybeans than upland
areas. However, the area of grass cover has decreased in both riparian and
upland areas between 2010 and 2020.

```
ggplot(data = cdl_chng) +
  geom_bar(aes(y = percarea,
```

```
               x = year,
                 fill = croptype),
              color = "black",
              position = "dodge",
              stat = "identity")   +
    scale_fill_manual(name = "Land Cover",
                      values = newcols) +
    facet_wrap(facets = vars(landtype),
                 ncol = 2) +
    labs(x = "Landscape Zone",
         y = "% Area") +
    theme_bw()
```

10.4   Summarizing Land Cover With Point Buffers

For this example, a set of twenty random points is generated within the North Deer Creek watershed. These could represent vegetation plots, wildlife sampling points, or the locations of other field-based measurements. Like the st_buffer() function, st_sample() generates an sfc object that must be converted to an sf object.

```
set.seed(32145)
plots <- st_sample(wshed_ndc, size = 20)
class(plots)
## [1] "sfc_POINT" "sfc"
plots <- st_as_sf(plots)
class(plots)
## [1] "sf"         "data.frame"
```

The st_sample() function uses a random number generator to produce x and y coordinates for the plot locations. The set.seed() function provides a starting seed for the random number generator so that it will produce the same results every time the code is run. If no random number seed is provided, it is derived from the current time in the computer's clock, and the results will be different every time the code is run.

The locations of these plots are plotted on top of the crop type data (Figure 10.11).

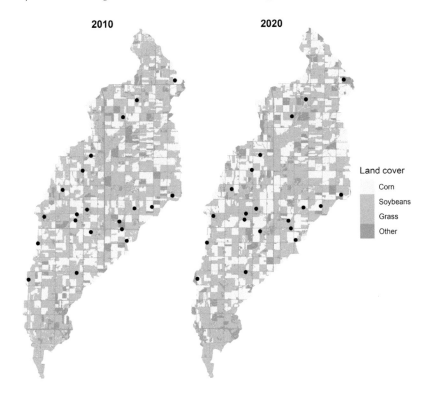

FIGURE 10.11
Plot locations overlaid on the crop type maps.

```
ggplot(data = cdl_df) +
  geom_raster(aes(x = x, y = y, fill = as.character(value))) +
  scale_fill_manual(name = "Land cover",
                    values = newcols2,
                    labels = newnames,
                    na.translate = FALSE) +
  geom_sf(data = plots, fill = NA) +
  facet_wrap(facets = vars(variable)) +
  coord_sf(expand = F) +
  theme_void() +
  theme(strip.text.x = element_text(size=12, face="bold"))
```

To summarize the percent cover of each land cover class within a buffer around each point, buffer polygons are created with a 200 m radius around each point.

```
st_geometry_type(plots, by_geometry = FALSE)
## [1] POINT
## 18 Levels: GEOMETRY POINT LINESTRING POLYGON ... TRIANGLE
plot_poly <- st_buffer(plots, dist = 200)
st_geometry_type(plot_poly, by_geometry = FALSE)
## [1] POLYGON
## 18 Levels: GEOMETRY POINT LINESTRING POLYGON ... TRIANGLE
```

To summarize the percent cover of each land cover class within a buffer around
each point, the segregate() function is used to convert the classified raster
data into a stack of indicator (1, 0) rasters. The proportion of each crop type
within each buffer is computed using the extract() function with the fun =
mean argument.

```
cdl_stk <- segregate(cdl_rc[["2020"]])
names(cdl_stk) <- newnames

plots_cdl <- extract(cdl_stk,
                     vect(plot_poly ),
                     fun = mean,
                     na.rm = T)
```

The resulting data frame has one row for each plot and one column for each
crop type.

```
plots_cdl
##    ID        Corn    Soybeans       Grass       Other
## 1   1 1.000000000 0.00000000 0.00000000 0.000000000
## 2   2 0.000000000 1.00000000 0.00000000 0.000000000
## 3   3 0.582733813 0.00000000 0.10791367 0.309352518
## 4   4 0.425531915 0.04255319 0.31914894 0.212765957
## 5   5 0.028571429 0.52142857 0.43571429 0.014285714
## 6   6 0.042857143 0.81428571 0.04285714 0.100000000
## 7   7 0.944055944 0.00000000 0.02797203 0.027972028
## 8   8 0.021276596 0.68085106 0.18439716 0.113475177
## 9   9 0.728571429 0.23571429 0.00000000 0.035714286
## 10 10 0.000000000 0.00000000 0.44927536 0.550724638
## 11 11 0.228571429 0.01428571 0.75000000 0.007142857
## 12 12 0.007142857 0.00000000 0.00000000 0.992857143
## 13 13 0.223076923 0.13076923 0.50000000 0.146153846
## 14 14 0.453237410 0.29496403 0.03597122 0.215827338
## 15 15 0.214285714 0.30000000 0.47142857 0.014285714
## 16 16 0.014285714 0.92857143 0.00000000 0.057142857
```

```
## 17 17 0.471014493 0.03623188 0.31159420 0.181159420
## 18 18 0.000000000 1.00000000 0.00000000 0.000000000
## 19 19 0.000000000 0.90647482 0.00000000 0.093525180
## 20 20 0.421428571 0.08571429 0.28571429 0.207142857
```

It is converted to a long format with all the crop type proportions in a single column and crop type names in a separate column.

```
plots_long <- plots_cdl %>%
  pivot_longer(Corn:Other,
               names_to = "croptype",
               values_to = "percarea")
plots_long
## # A tibble: 80 x 3
##       ID croptype percarea
##    <dbl> <chr>       <dbl>
## 1      1 Corn            1
## 2      1 Soybeans        0
## 3      1 Grass           0
## 4      1 Other           0
## 5      2 Corn            0
## 6      2 Soybeans        1
## 7      2 Grass           0
## 8      2 Other           0
## 9      3 Corn        0.583
## 10     3 Soybeans        0
## # ... with 70 more rows
```

The distribution of crop type proportions within the plot buffer can be visualized as a boxplot (Figure 10.12).

```
ggplot(data = plots_long) +
  geom_boxplot(aes(x = croptype,
                   y = percarea)) +
  xlab("Land Cover Class") +
  ylab("Cover Proportion") +
  theme_bw()
```

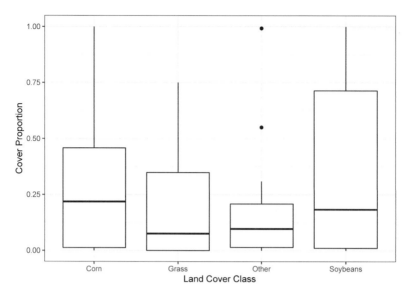

FIGURE 10.12
Percentage of crop types sampled within plot buffers.

10.5   Practice

1. Repeat the stream buffer analysis for the Hidewood Creek watershed using a 200 m stream buffer width.

2. Generate 10 random plot locations in the North Deer Creek watershed using a different random number seed (12345). Reclassify the CDL data into three classes: 1) Corn plus soybean, 2) grass, and 3) all other crop types and generate a boxplot of their distributions within 300 m buffers around the plots.

11

Application—Wildfire Severity Analysis

This chapter addresses the application of satellite remote sensing data for mapping wildfire severity patterns and analyzing their associations with the wildland-urban interface and topography. Fire severity is defined as the level of impact that a fire has on the ecosystem that was burned. In forest ecosystems, major aspects of severity include the death of overstory trees and combustion of soil organic matter. One way to measure fire severity is to obtain remotely-sensed imagery before and after the fire and assess the differences in the images to measure how much change in vegetation and soils has occurred after the fire (Miller and Thode, 2007). These data can be combined with other social and environmental datasets to study the determinants and effects of fire severity. Understanding these relationships is important for developing management strategies to sustain the beneficial ecological effects of fire while reducing the negative impacts on people and property.

In this chapter, wildfire data will be analyzed by applying the techniques covered in early chapters and introducing a new statistical modeling technique to characterize the influences of environmental factors on spatial patterns of fire severity. Most of the R packages needed to conduct these analyses have been introduced in the previous chapter. In addition, the **mgcv** package (Wood, 2022) will be used for generalized additive modeling, and the **visreg** package (Breheny and Burchett, 2020) will be used to help visualize the model results.

```
library(tidyverse)
library(terra)
library(ggspatial)
library(sf)
library(cowplot)
library(mgcv)
library(visreg)
source("rasterdf.R")
```

11.1   Remote Sensing Image Analysis

We will use Landsat imagery to develop a map of fire severity for the High Park fire, which burned more than 87,000 acres near Fort Collins, CO in 2012. A series of Landsat satellites have provided high-resolution (30 m) multispectral imagery beginning with the launch of Landsat 4 in 1984. The data for this analysis were obtained from the Monitoring Trends in Burn Severity (MTBS) project (https://www.mtbs.gov/). This project uses satellite remote sensing to produce maps of fire severity for every major fire that occurs in the United States (Eidenshink et al., 2007). For illustration, this chapter will start with the raw satellite imagery provided by MTBS and show how it is processed to develop an index of burn severity. To start, two Landsat images are imported. One was collected in 2011 before the fire, and one was collected in 2013 after the fire.

```
landsat_pre <- rast("co4058910540420120609_20110912_l5_refl.tif")
landsat_post <- rast("co4058910540420120609_20130901_l8_refl.tif")
```

Each of the resulting SpatRaster objects is a raster of 30 m square cells with 850 rows by 1149 columns. The landsat_pre object contains Landsat 5 data with six spectral bands.

```
nrow(landsat_pre)
## [1] 850
ncol(landsat_pre)
## [1] 1149
res(landsat_pre)
## [1] 30 30
nlyr(landsat_pre)
## [1] 6
ext(landsat_pre)[1:4]
##     xmin     xmax     ymin     ymax
## -801045 -766575 1985745 2011245
```

The landsat_post object contains Landsat 8 data with eight spectral bands. The first six bands of each layer match up, and the seventh and eighth bands in landsat_post contain two additional bands that we will not use here.

```
nrow(landsat_post)
## [1] 850
ncol(landsat_post)
## [1] 1149
```

FIGURE 11.1
False-color image of 2011 pre-fire vegetation condition on the High Park fire.

```
res(landsat_post)
## [1] 30 30
nlyr(landsat_post)
## [1] 8
ext(landsat_post)[1:4]
##    xmin    xmax    ymin    ymax
## -801045 -766575 1985745 2011245
```

In general, R is not as powerful for visualizing and exploring raw remote sensing imagery as dedicated remote sensing software. However, it is helpful to be able to quickly view remotely sensed images in R. The plotRGB() function in the **terra** package displays the Landsat images as false color composites, with the short-wave infrared band (layer 6) displayed as red, near-infrared (layer 4) displayed as green, and green (layer 2) displayed as blue. The pre-fire image is dominated by healthy vegetation, which reflects near-infrared radiation and appears green in the false-color image (Figure 11.1).

```
plotRGB(landsat_pre,
        r = 6,
        g = 4,
        b = 2)
```

FIGURE 11.2
False-color image of 2013 post-fire vegetation condition on the High Park fire.

Large areas of trees were killed by the fire in 2012, and soils in the burned areas were blackened by char. These areas reflect less near-infrared radiation and more short-wave infrared radiation. As a result, much of the false color image appears reddish brown in 2013 after the fire (Figure 11.2).

```
plotRGB(landsat_post,
        r = 6,
        g = 4,
        b = 2)
```

Pre- and post-fire landscape conditions are characterized using the normalized burn ratio (NBR) index. This index measures the contrast between the near-infrared band (layer 4), which is highest in areas of healthy green vegetation, and a shortwave-infrared band (layer 6), which is highest in unvegetated areas with blackened or exposed soils. Thus, the NBR is highest in areas with undisturbed forest vegetation and lowest in areas where fire has killed most or all of the vegetation and blackened the soil. The differenced normalized burn index (DNBR) is calculated as the difference between pre- and post-fire NBR, providing an estimate of the effects of the fire on vegetation and soils.

```
nbr_pre <- 1000 * (landsat_pre[[4]] - landsat_pre[[6]]) /
  (landsat_pre[[4]] + landsat_pre[[6]])
```

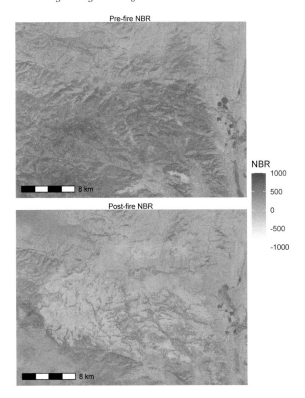

FIGURE 11.3
Pre- and post-fire NBR indices for the High Park fire.

```
nbr_post <- 1000 * (landsat_post[[4]] - landsat_post[[6]]) /
  (landsat_post[[4]] + landsat_post[[6]])
dnbr <- nbr_pre - nbr_post
```

The pre-fire and post-fire NBR rasters are combined into a multilayer raster object and mapped (Figure 11.3). A scale bar is also added for reference. Because increasing levels of NBR are generally associated with increasing tree cover, a light yellow to the dark green color ramp is used. There are more areas with low NBR after the fire, but some locations already had low NBR before the fire.

```
nbr_stack <- c(nbr_pre, nbr_post)
names(nbr_stack) <- c("Pre-fire NBR", "Post-fire NBR")
nbr_stack_df <- rasterdf(nbr_stack)

ggplot(nbr_stack_df) +
```

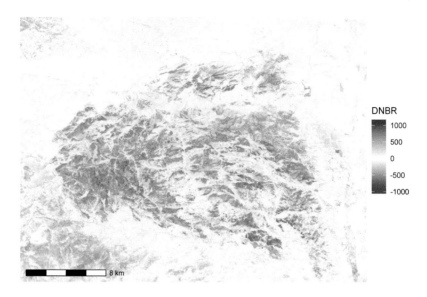

FIGURE 11.4
DNBR index for the High Park fire.

```
geom_raster(aes(x = x,
                y = y,
                fill = value)) +
scale_fill_gradient(name = "NBR",
                    low = "lightyellow",
                    high = "darkgreen") +
coord_sf(expand = FALSE) +
annotation_scale(location = 'bl') +
facet_wrap(facets = vars(variable),
           ncol = 1) +
theme_void()
```

The DNBR index highlights locations where vegetation changed after the
fire (Figure 11.4). DNBR is centered at zero, with positive values indicating
decreases in NBR and negative values indicating increases. Therefore, the
scale_fill_gradient2() is used to create a bicolor ramp centered at zero.

```
dnbr_df <- rasterdf(dnbr)

ggplot(dnbr_df) +
  geom_raster(aes(x = x,
```

```
                   y = y,
                   fill = value)) +
  scale_fill_gradient2(name = "DNBR",
                       low = "blue",
                       high = "red",
                       midpoint = 0) +
  coord_sf(expand = F) +
  annotation_scale(location = 'bl') +
  theme_void()
```

11.2   Burn Severity Classification

The DNBR index will be used to classify the image into discrete levels of
burn severity. The classify() function is used to assign severity classes based
on DNBR. Previous applications of classify() with land cover data used
one-to-one lookup tables to assign each raster value to a new class. Because
DNBR is a continuous variable, it is necessary to use a lookup table based on
ranges of values.

The easiest way to do this is to create a matrix with three columns: the first
column contains the lower value of each range, the second column contains the
upper value of each range, and the third column contains the new value to be
assigned to the range. The matrix() function takes a vector of data, and the
ncol = 3 and byrow = TRUE arguments indicate that the first three values will
be the first row, the second three values with be the second row, etc. Then,
the classify() function is used with the raster dataset to be reclassified as
the first argument and the lookup table as the second argument.

```
rclas <- matrix(c(-Inf, -970, NA,   # Missing data
                  -970, -100, 5,    # Increased greenness
                  -100, 80, 1,      # Unburned
                  80, 265, 2,       # Low severity
                  265, 490, 3,      # Moderate severity
                  490, Inf, 4),     # High severity
                  ncol = 3, byrow = T)

severity <- classify(dnbr, rclas)
```

Inferring fire severity from vegetation change is based on the assumption that
the observed changes are the result of the wildfire and not other drivers of land
cover and land use change. Therefore, it is advisable to mask the fire severity

FIGURE 11.5
Severity classes for the High Park fire.

raster to the known perimeter of the wildfire. This can be accomplished by
reading in the fire boundary, rasterizing it to match the classified fire severity
raster, and masking the areas outside of the fire perimeter.

```
fire_bndy <- st_read("co4058910540420120609_20110912_20130901_bndy.shp",
                     quiet = TRUE)
bndy_rast <- rasterize(vect(fire_bndy),
                       severity,
                       field = "Event_ID")
severity <- mask(severity, bndy_rast)
```

Vectors of colors and names for the five severity classes are specified and are
used with the masked fire severity dataset to generate a map (Figure 11.5).
High severity indicates areas where crown fire or high-intensity surface fire
killed most of the trees. Low severity indicates areas with low-intensity surface
fires where most trees survived. Mixed severity indicates an intermediate level
of tree survival. In some cases, rapid postfire growth can occur where there
was sparse vegetation prior to the fire and burning triggers a flush of grasses
and other herbaceous vegetation growth.

```
SCcolors = c("darkgreen",
             "cyan3",
```

```
            "yellow",
            "red",
            "green")
SCnames = c("Unburned",
            "Low",
            "Moderate",
            "High",
            "> Green")
severity_df <- rasterdf(severity)

ggplot(severity_df) +
  geom_raster(aes(x = x,
                  y = y,
                  fill = as.character(value))) +
  scale_fill_manual(name = "Severity Class",
                    values = SCcolors,
                    labels = SCnames,
                    na.translate = FALSE) +
  annotation_scale(location = 'bl') +
  coord_fixed(expand = F) +
  theme_void()
```

11.3   The Wildland-Urban Interface

We will analyze these burn severity patterns by overlaying them with several
other geospatial datasets. The wildland-urban interface (WUI) is defined as
the zone where houses and other infrastructure are location in close vicinity
to wildland vegetation. The "intermix" is characterized by houses and other
buildings scattered throughout wildlands at low densities, whereas the "interface" is the zone where denser urban settlements are adjacent to wildland
vegetation (Radeloff et al., 2005). The WUI is a significant concern for fire
management because of the difficulty and expense of protecting large numbers
of structures, particularly when they are dispersed over large areas of intermix
WUI. Thus, there is interest in knowing the degree to which large wildfires
like the High Park fire are occurring within or close to the WUI.

The WUI can be identified via geospatial analysis that overlays vegetation
data from the NLCD with housing density data from the US Census. WUI
data for the United States can be downloaded on a state-by-state basis from
http://silvis.forest.wisc.edu/data/wui-change/. Because of the large

sizes of these statewide datasets, this example uses a smaller version that has already been clipped to a portion of Colorado.

```
wui <- st_read("co_wui_cp12_clip.shp", quiet=TRUE)
```

The WUI dataset will need to be rasterized and cropped to match the geometric characteristics of the fire severity dataset. Both datasets are in a similar Albers Equal Area coordinate system, although there is a minor difference in the definition of the spheroid.

```
writeLines(st_crs(wui)$WktPretty)
## PROJCS["NAD_1983_Albers",
##      GEOGCS["NAD83",
##          DATUM["North_American_Datum_1983",
##              SPHEROID["GRS 1980",6378137,298.257222101],
##              AUTHORITY["EPSG","6269"]],
##          PRIMEM["Greenwich",0],
##          UNIT["Degree",0.0174532925199433]],
##      PROJECTION["Albers_Conic_Equal_Area"],
##      PARAMETER["latitude_of_center",23],
##      PARAMETER["longitude_of_center",-96],
##      PARAMETER["standard_parallel_1",29.5],
##      PARAMETER["standard_parallel_2",45.5],
##      PARAMETER["false_easting",0],
##      PARAMETER["false_northing",0],
##      UNIT["metre",1,
##          AUTHORITY["EPSG","9001"]],
##      AXIS["Easting",EAST],
##      AXIS["Northing",NORTH]]
writeLines(st_crs(severity)$WktPretty)
## PROJCS["USA_Contiguous_Albers_Equal_Area_Conic_USGS_version",
##      GEOGCS["NAD83",
##          DATUM["North_American_Datum_1983",
##              SPHEROID["GRS 1980",6378137,298.257222101004]],
##          PRIMEM["Greenwich",0],
##          UNIT["degree",0.0174532925199433,
##              AUTHORITY["EPSG","9122"]],
##          AUTHORITY["EPSG","4269"]],
##      PROJECTION["Albers_Conic_Equal_Area"],
##      PARAMETER["latitude_of_center",23],
##      PARAMETER["longitude_of_center",-96],
##      PARAMETER["standard_parallel_1",29.5],
##      PARAMETER["standard_parallel_2",45.5],
##      PARAMETER["false_easting",0],
```

FIGURE 11.6
Vector data characterizing WUI classes for the High Park fire.

```
##      PARAMETER["false_northing",0],
##      UNIT["metre",1,
##          AUTHORITY["EPSG","9001"]],
##      AXIS["Easting",EAST],
##      AXIS["Northing",NORTH]]
```

The WUI vector data are first reprojected to match the coordinate system of the fire severity dataset and are then cropped to its boundary. The map shows the WUI polygons, which correspond to U.S. Census blocks, in the different WUI classes (Figure 11.6).

```
wui_reproj <- st_transform(wui, crs(severity))
wui_crop <- st_crop(wui_reproj, severity)

ggplot(data = wui_crop) +
  geom_sf(aes(fill = as.character(WUIFLAG10))) +
  scale_fill_manual(name = "WUI Class",
                    values = c("Gray", "Orange", "Red"),
                    labels = c("Non-WUI", "Intermix", "Interface"),
                    na.translate = FALSE) +
  coord_sf(expand = FALSE) +
  theme_void()
```

FIGURE 11.7
Raster data characterizing WUI classes for the High Park fire.

Next, the `rasterize()` function is used to convert the vector WUI dataset
into a raster layer with the same geometric characteristics as the fire severity
layer. The WUIFLAG10 field provides the values for the output raster: 0 =
Non-WUI, 1 = Intermix, and 2 = Interface (Figure 11.7).

```
wui_rast <- rasterize(vect(wui_crop),
                      severity,
                      field = "WUIFLAG10")
wui_rast_df <- rasterdf(wui_rast)

ggplot(wui_rast_df) +
  geom_raster(aes(x = x,
                  y = y,
                  fill = as.character(value))) +
  scale_fill_manual(name = "WUI Class",
                    values = c("Gray", "Orange", "Red"),
                    labels = c("Non-WUI", "Intermix", "Interface"),
                    na.translate = FALSE) +
  coord_sf(expand = FALSE) +
  theme_void()
```

Distance from the WUI is calculated using the `distance()` function. Distance
is calculated for all cells with an `NA` value to the nearest cell that does not have
an `NA` value. Therefore, the WUI raster dataset must first be reclassified into

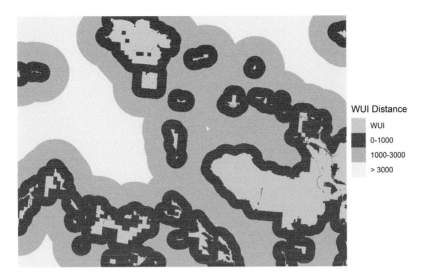

FIGURE 11.8
Raster data distance from the nearest WUI for the High Park fire.

a raster where all WUI cells have a value and all non-WUI cells are NA. This type of simple conditional assignment can be made using the ifel() function, which takes a logical expression as the first argument, the value to assign if the expression is TRUE as the second argument, and the value to assign if the expression if FALSE as the third argument.

```
wui_na <- ifel(wui_rast == 0, NA, 1)
wui_dist <- distance(wui_na)
```

The raster containing continuous distance values is then reclassified with the classify() function into a discrete raster with four distance classes using the same approach applied previously for DNBR. The distance to WUI classes is displayed as a categorical map (Figure 11.8).

```
rclas <- matrix(c(-Inf, 0, 1,
                  0, 1000, 2,
                  1000, 3000, 3,
                  3000, Inf, 4),
               ncol = 3, byrow = T)
wui_rcls <- classify(wui_dist, rcl = rclas)
wui_rcls_df <- rasterdf(wui_rcls)

ggplot(wui_rcls_df) +
```

```
geom_raster(aes(x = x,
                y = y,
                fill = as.factor(value))) +
scale_fill_manual(name = "WUI Distance",
                  values = c("Gray",
                             "Red",
                             "Orange",
                             "Yellow"),
                  labels = c("WUI",
                             "0-1000",
                             "1000-3000",
                             "> 3000"),
                  na.translate = FALSE) +
coord_sf(expand = FALSE) +
theme_void()
```

The `crosstab()` function is used to calculate the distribution of fire severity classes for every WUI distance class. The output is then manipulated to give the columns shorter names, convert counts of 30 m square cells to hectares, and convert the WUI distance and severity variables to labeled factors.

```
wui_xtab <- crosstab(c(wui_rcls,
                       severity))
wui_df <- as_tibble(wui_xtab)

wui_df <- wui_df %>%
  rename(wuidist = 1, sev = 2, ha = 3) %>%
  mutate(ha = ha * 900 / 10000,
         wuidist = factor(wuidist,
                          levels = 1:4,
                          labels = c("WUI",
                                     "0-1000",
                                     "1000-3000",
                                     "> 3000")),
         sev = factor(sev,
                      levels = 1:5,
                      labels = SCnames))
```

A bar chart comparing these categories shows that there is an association between distance to the WUI and fire severity (Figure 11.9). Inside the WUI and close to the WUI, unburned and low severity are the most common severity classes. However, the relative amounts of moderate and high severity classes increase with distance from the WUI, and at the farthest distance, high severity is the most abundant class.

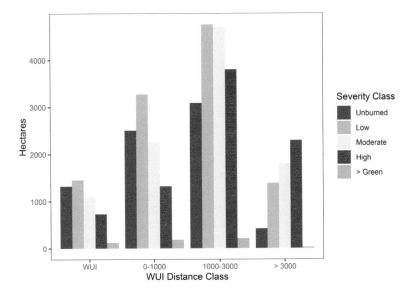

FIGURE 11.9
Raster data distance from the nearest WUI for the High Park fire.

```
ggplot(data = wui_df) +
  geom_bar(aes(x = wuidist,
               y = ha,
               fill = sev),
           position = "dodge",
           stat = "identity") +
  scale_fill_manual(name = "Severity Class",
                    values = SCcolors) +
  labs(x = "WUI Distance Class",
       y = "Hectares") +
  theme_bw()
```

11.4 Topographic Effects

The next analysis will explore the relationship between the observed pattern of fire severity and topography. Elevation is often related to fire severity because the climate is strongly influenced by elevation, and different forest vegetation types and fire regimes are found at different elevations as a result. In the Colorado Front range, higher-elevation forests are often dominated by lodgepole pine (*Pinus contorta*), which occurs in dense stands that are susceptible to high-severity crown fires. Lower elevations are dominated by other tree species such as Douglas-fir (*Pseudotsuga menziesii*) and ponderosa pine (*Pinus ponderosa*) that are more like to survive when an area is burned. Wildfires are also sensitive to slope angle, spreading more rapidly up and more slowly down steep slopes. South-facing slopes receive more direct solar radiation than north-facing slopes. The resulting drier conditions affect the vegetation community and reduce fuel moisture, which can lead to higher fire severity on south-facing slopes.

11.4.1 Data processing

The elevation data were obtained from the National Elevation Dataset (NED) through the United States Geological Survey's National Map (https://www.usgs.gov/the-national-map-data-delivery). These data are in a geographic projection and must be reprojected to match the fire severity data. By providing the `severity` raster as the second argument to the `project()` function, the output is projected to the same CRS and also matched to the raster geometry (origin, cell size, and extent) of the existing fire severity raster.

```
elev <- rast("USGS_1_n41w106_20220331.tif")
writeLines(st_crs(elev)$WktPretty)
## GEOGCS["NAD83",
##      DATUM["North_American_Datum_1983",
##          SPHEROID["GRS 1980",6378137,298.257222101004]],
##      PRIMEM["Greenwich",0],
##      UNIT["degree",0.0174532925199433,
##          AUTHORITY["EPSG","9122"]],
##      AXIS["Latitude",NORTH],
##      AXIS["Longitude",EAST],
##      AUTHORITY["EPSG","4269"]]
elev_crop <- project(elev,
                     severity,
                     method = "bilinear")
```

The map shows that elevation in the Colorado Front Range generally increases from east to west and that there are several drainages and ridgelines within the boundary of the High Park fire (Figure 11.10).

```
elev_crop_df <- rasterdf(elev_crop)
ggplot(elev_crop_df) +
  geom_raster(aes(x = x, y = y, fill = value)) +
  scale_fill_distiller(name = "Elevation (m)",
                       palette = "Oranges") +
```

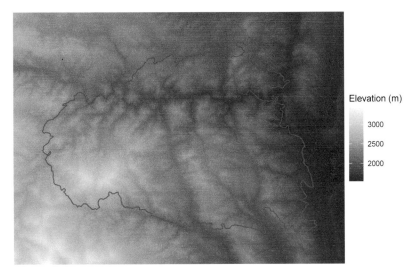

FIGURE 11.10
Elevation within the boundary of the High Park fire.

```
geom_sf(data = fire_bndy, fill = NA) +
coord_sf(expand = FALSE) +
theme_void()
```

Slope angle can be calculated using the `terrain()` function and then mapped (Figure 11.11).

```
slopedeg <- terrain(elev_crop,
                    v="slope",
                    unit="degrees")
slopedeg_df <- rasterdf(slopedeg)
ggplot(slopedeg_df) +
  geom_raster(aes(x = x, y = y, fill = value)) +
  scale_fill_distiller(name = "Slope (degrees)",
                       palette = "Oranges") +
  geom_sf(data = fire_bndy, fill = NA) +
  coord_sf(expand = FALSE) +
  theme_void()
```

Slope aspect is similarly calculated using the `terrain()` function with units as radians. A north aspect will have a value of 0, an east aspect a value of 0.5π, a south aspect a value of π, and a west aspect a value of 1.5π. The `cos()` function is then used to convert the circular aspect angle into a linear

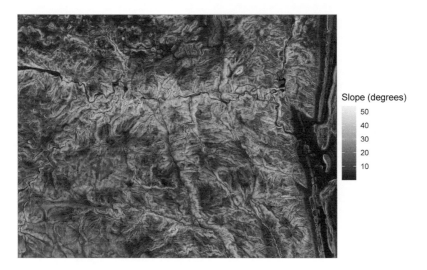

FIGURE 11.11
Slope angle within the boundary of the High Park fire.

north-south index, where values of 1 indicate north-facing slopes and values of −1 indicate south-facing slopes (Figure 11.12).

```
aspect <- terrain(elev_crop, v="aspect", unit="radians")
nsaspect <- cos(aspect)
nsaspect_df <- rasterdf(nsaspect)

ggplot(nsaspect_df) +
  geom_raster(aes(x = x, y = y, fill = value)) +
  scale_fill_distiller(name = "Aspect (N-S index)",
                       palette = "Oranges") +
  geom_sf(data = fire_bndy, fill = NA) +
  coord_sf(expand = FALSE) +
  theme_void()
```

The sin() function is similarly used to create an east-west index where east-facing slopes have a value of 1 and west-facing slopes have a value of −1 (Figure 11.13).

```
ewaspect <- sin(aspect)
ewaspect_df <- rasterdf(ewaspect)

ggplot(ewaspect_df) +
  geom_raster(aes(x = x, y = y, fill = value)) +
```

FIGURE 11.12
North-south aspect index within the boundary of the High Park fire.

FIGURE 11.13
East-west aspect index within the boundary of the High Park fire.

```
scale_fill_distiller(name = "Aspect (E-W index)",
                     palette = "Oranges") +
geom_sf(data = fire_bndy, fill = NA) +
coord_sf(expand = FALSE) +
theme_void()
```

To analyze the relationships between fire severity and these topographic indices, a sample of points inside the boundary of the High Park fire is needed. The first step is to combine DNBR with the topographic indices into a multilayer raster object.

```
fire_stack <- c(dnbr,
                elev_crop,
                slopedeg,
                nsaspect,
                ewaspect)
```

The `st_sample()` function is used to generate a set of random points as was done in Chapter 10. In this case, several additional arguments are provided. The points are constrained to fall inside the `fire_bndy` polygon data set. The `type = SSI` argument indicates that a simple sequential inhibition process will be used to generate the points. After the first random point is generated, subsequent points are only accepted if they are farther than a threshold distance from the nearest point. The `r` argument indicates the minimum distance between points, and `n` indicates the number of points to sample. The resulting `sample_pts` dataset is in the same coordinate reference system as `fire_bndy`, but the coordinate reference system is not automatically defined and must be specified manually.

```
set.seed(23456)
sample_pts <- st_sample(
  fire_bndy,
  type = "SSI",
  r = 500,
  n = 300
)
st_crs(sample_pts)
## Coordinate Reference System: NA
st_crs(sample_pts) <- st_crs(fire_bndy)
```

11.4.2   Generalized additive modeling

Setting a minimum distance is desirable in this case because the values of nearby raster cells tend to be strongly correlated with one another, and using widely-spaced points help to ensure broad coverage and maximize the independence of each sample (Figure 11.14). As discussed in Chapter 10, setting a random number seed ensures that running the code multiple times will generate the same sequence of random numbers.

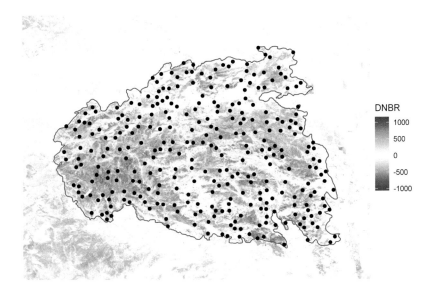

FIGURE 11.14
Sample points and fire boundary overlaid on the DNBR index for the High
Park fire.

```
ggplot(dnbr_df) +
  geom_raster(aes(x = x,
                  y = y,
                  fill = value)) +
  scale_fill_gradient2(name = "DNBR",
                       low = "blue",
                       high = "red",
                       midpoint = 0) +
  geom_sf(data = fire_bndy, fill = NA) +
  geom_sf(data = sample_pts) +
  coord_sf(expand = F) +
  theme_void()
```

The `extract()` function is then used to extract raster data at these point
locations, and the columns of the resulting data frame are renamed.

```
fire_pts <- extract(fire_stack, vect(sample_pts))

fire_pts <- rename(fire_pts,
                   dnbr = 2,
```

```
        elevation = 3,
        slope = 4,
        nsaspect = 5,
        ewaspect = 6)
```

To explore the relationships between fire severity and topography, a generalized additive model (GAM) is applied using the `gam()` function from the **mgcv** library. The specification of the model is similar to the linear models that were previously run using the `lm()` function. A model formula is specified with dependent variables on the left side of the tilde (~) and independent variables separated by the plus (+) symbol on the right side. In addition, each of the independent variables is wrapped in the `s()` function, which by default models the dependent variable as a smooth thin-plate spline function. This approach is warranted when the underlying relationships are not assumed to be linear.

```
fire_gam <- gam(dnbr ~
                s(elevation) +
                s(slope) +
                s(nsaspect) +
                s(ewaspect),
              data = fire_pts)
```

The summary method for a `gam` object looks generally similar to that of an `lm()` object, but upon close inspection, has some important differences. For example, there is a table of test statistics and p-values for each of the independent variables, but there are no coefficient values as with a linear model. The estimated degrees of freedom (`edf`) provides information about the degree of non-linearity in the relationships, with higher values indicating more complex non-linear relationships.

```
class(fire_gam)
## [1] "gam" "glm" "lm"
summary(fire_gam)
##
## Family: gaussian
## Link function: identity
##
## Formula:
## dnbr ~ s(elevation) + s(slope) + s(nsaspect) + s(ewaspect)
##
## Parametric coefficients:
##             Estimate Std. Error t value Pr(>|t|)
## (Intercept)   272.54      10.23   26.63   <2e-16 ***
```

```
## ---
## Signif. codes:
## 0 '***' 0.001 '**' 0.01 '*' 0.05 '.' 0.1 ' ' 1
##
## Approximate significance of smooth terms:
##                   edf Ref.df      F  p-value
## s(elevation) 5.489  6.656 17.609  < 2e-16 ***
## s(slope)     2.708  3.418  9.257 4.21e-06 ***
## s(nsaspect)  1.000  1.000 53.578  < 2e-16 ***
## s(ewaspect)  4.351  5.323  1.547   0.167
## ---
## Signif. codes:
## 0 '***' 0.001 '**' 0.01 '*' 0.05 '.' 0.1 ' ' 1
##
## R-sq.(adj) =  0.411   Deviance explained = 43.8%
## GCV =  33017  Scale est. = 31416    n = 300
```

To understand the smoothed relationships modeled by a GAM, they need to be plotted. The `gam` object has a `plot()` method that uses base R graphics to graph the partial residuals as a function of each independent variable. These partial residuals represent the relationships between DNBR and each topographic index after removing the modeled effects of all the other topographic indices (Figure 11.15). The output shows that DNBR has a unimodal relationship with an elevation that peaks at approximately 2700 meters. The relationships with slope are also unimodal with a peak around 22 degrees. The relationships with the north-south index are linear, with fire severity highest on south-facing aspects. The relationship with the east-west aspect is comparatively weak and does not show a clear pattern.

```
plot(fire_gam, pages = 1)
```

Partial regression plots can also be generated using the `visreg()` function from the **visreg** package. By specifying the `gg = TRUE` argument, the plots are generated using `ggplot()`, and additional ggplot functions can be specified to modify their appearance. Here, a partial plot for each independent variable is generated and saved to a ggplot object.

```
elev_gg <- visreg(fire_gam,
                  "elevation",
                  gg = TRUE) +
  theme_bw()
slope_gg <- visreg(fire_gam,
                   "slope",
```

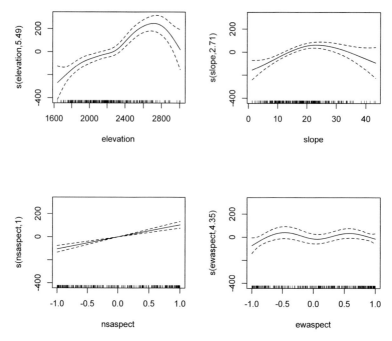

FIGURE 11.15
Partial regression plots for the relationships between topographic indices and
fire severity.

```
                        gg = TRUE) +
    theme_bw()
nsasp_gg <- visreg(fire_gam,
                       "nsaspect",
                       gg = TRUE) +
    theme_bw()
ewasp_gg <- visreg(fire_gam,
                       "ewaspect",
                       gg = TRUE) +
    theme_bw()
```

Then, as was shown in Chapter 5, the plot_grid() function from the **cowplot**
package can be used to arrange these multiple plots into a larger multipanel
graphic (Figure 11.16).

```
plot_grid(elev_gg, slope_gg, nsasp_gg, ewasp_gg,
          labels = c("A)", "B)", "C)", "D)",
                     label_size = 12),
```

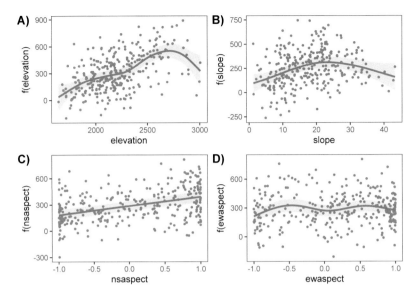

FIGURE 11.16
Partial regression plots for the relationships between topographic indices and
fire severity generated using the cowplot package.

```
ncol = 2,
hjust = 0,
label_x = 0,
align = "hv")
```

11.5 Practice

1. One of the reasons why fire severity was lowest near the WUI may
   have been that the WUI was concentrated at lower elevations where
   the forest vegetation is often resilient to fire. Compare these two
   variables by classifying elevation into four categories and generating
   a plot (Figure 11.9) that compares elevation with distance from the
   WUI.

2. Another metric of burn severity is the relative differenced normalized burn ratio (RDNBR), which normalizes the DNBR by divid-
   ing it by the pre-fire NBR. The RDNBR is calculated as `rdnbr =
   dnbr/sqrt(abs(nbr_pre))`. Calculate the RDNBR for the High Park

fire and generate a composite figure that includes both DNBR and RDNBR mapped as continuous variables. Compare the maps and assess the similarities and differences between the two indices.

3. Select and download burn severity data for another wildfire from the MTBS website (`https://www.mtbs.gov/`) along with elevation data for the same area from the National Map (`https://www.us gs.gov/the-national-map-data-delivery`). Repeat the analysis of topographic effects on burn severity described in this chapter and see if you obtain similar results in a different location.

12

Application—Species Distribution Modeling

Climate is one of the most important factors influencing the geographic ranges of species. Each species is adapted to a particular set of climate conditions, which is referred to as its *climate niche*. The *fundamental niche* is the hypothetical range of conditions in which a species could persist independently of other species. The *realized niche* is the narrower range of conditions in which a species can persist when interactions with other species, such as predation and competition, are also taken into account. The realized niche can be characterized by observing where a species is present and absent over large areas and modeling the association between species occurrence and climate variables measured where the species is present and absent (Elith and Leathwick, 2009).

Because of the changes in Earth's climate that are occurring as a result of anthropogenic greenhouse gas emissions, the locations that are most climatically favorable for different species will shift over time. These changes have the potential to radically alter local communities and ecosystems. There is concern that some species may be threatened if they are not able to migrate fast enough to keep up with shifting climate patterns or become disjunct from other mutualistic species such as pollinators. This chapter will explore these issues by modeling the relationships between tree species and climate in mountains of the Pacific Northwest and analyzing how species distributions may change under a future climate change scenario.

Three new packages will be used in this chapter. The **spatstat** package (Baddeley et al., 2022) implements methods for the manipulation and analysis of point pattern data. The **dismo** package (Hijmans et al., 2022) contains a suite of tools for species distribution modeling. The **ROCR** package (Sing et al., 2005) is used to generate receiver operating characteristics curves and other graphs and statistics to assess prediction accuracy.

```
library(tidyverse)
library(sf)
library(terra)
library(colorspace)
library(ggspatial)
library(dismo)
```

DOI: 10.1201/9781003326199-12

```
library(spatstat)
library(ROCR)
source("rasterdf.R")
```

12.1   Tree Species Data

The study will encompass the Cascade Mountain Range in Washington State. This area includes three ecoregions: the Cascades, East Cascades, and North Cascades. These polygons are read in from a shapefile. Two additional datasets are read in with presence and absence points for subalpine fir (*Abies lasiocarpa*) and Douglas-fir (*Pseudotsuga menziesii*). Note that Douglas is a proper name, and Douglas-fir is not a true fir in the genus *Abies*, which is why the common name of the species is capitalized and hyphenated. The data are from forest inventory plots monitored by the U.S. Forest Service and other land management agencies. These data files were derived from the archived datasets provided by Charney et al. (2021).

```
wacascades <- st_read("wacascades.shp", quiet = TRUE)
abla <- read_csv("abla.csv")
psme <- read_csv("psme.csv")
```

The `abla` and `psme` data frames contain columns for longitude and latitude (`X` and `Y`) along with a column of binary (1/0) data for species presence/absence. These data frames are converted into `sf` objects and assigned the correct coordinate system (geographic with NAD83 datum).

```
abla_pts <- st_as_sf(abla,
                     coords = c("X", "Y"))
st_crs(abla_pts) <- 4326
psme_pts <- st_as_sf(psme,
                     coords = c("X", "Y"))
st_crs(psme_pts) <- 4326
```

The map of presence–absence points shows that subalpine fir is restricted to relatively high elevations near the Cascade crest (Figure 12.1).

```
ggplot() +
  geom_sf(data = abla_pts,
          aes(color = as.character(abla)),
```

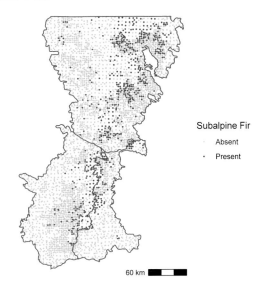

FIGURE 12.1
Occurrence of subalpine fir in forest inventory plots in the Washington
Cascades.

```
            size = 0.25) +
  geom_sf(data = wacascades,
            fill = NA) +
  scale_color_manual(name = "Subalpine Fir",
                     values = c("gray", "darkgreen"),
                     labels = c("Absent", "Present")) +
  annotation_scale(location = 'br') +
  theme_void()
```

In contrast, Douglas-fir is distributed more widely at lower elevations and
along the eastern edge of the Cascades (Figure 12.2).

```
ggplot() +
  geom_sf(data = psme_pts,
          aes(color = as.character(psme)),
          size = 0.25) +
  geom_sf(data = wacascades,
          fill = NA) +
  scale_color_manual(name = "Douglas-fir",
                     values = c("gray", "darkgreen"),
                     labels = c("Absent", "Present")) +
```

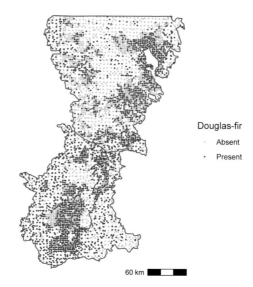

FIGURE 12.2
Occurrence of Douglas-fir in forest inventory plots in the Washington Cascades.

```
annotation_scale(location = 'br') +
theme_void()
```

One of the challenges with visualizing the points is that the overall density of plots is variable. In general, there is a higher density of plots on public lands such as National Forests and a lower density on private lands. One way to try to see the occurrence patterns more clearly is to compute kernel-weighted local means. This technique is a type of point-to-raster conversion in which each raster cell in the output is assigned a mean of the local presence/absence values, where points located closer to the cell are given a higher weight than those more distant.

To make this transformation, the point data must be in a projected coordinate system instead of a geographic coordinate system. The Cascades county shapefile as well as the subalpine fir and Dougas-fir point datasets are all reprojected into UTM zone 10 north.

```
boundary_utm <- st_transform(wacascades, 32610)
abla_utm <- st_transform(abla_pts, 32610)
psme_utm <- st_transform(psme_pts, 32610)
```

The **spatstat** package contains a function for generating kernel-weighted means. However, it cannot directly read the data as sf objects. They must first be converted to **spatstat** ppp objects using the as.ppp() function. The column of ones and zeroes is automatically read in and stored as the "marks" for the point pattern object. Then the as.owin() function is used to create an owin (observation window) object and assign it to the ppp object.

```
abla_ppp <- as.ppp(abla_utm)
class(abla_ppp)
## [1] "ppp"
Window(abla_ppp) <- as.owin(boundary_utm)
abla_ppp
## Marked planar point pattern: 4356 points
## marks are numeric, of storage type  'double'
## window: polygonal boundary
## enclosing rectangle: [511678.3, 760623.9] x [5046278,
## 5432695] units
```

The Smooth.ppp() function can now be used to compute kernel-weighted means. The sigma argument specifies the bandwidth of the kernel, which can be conceptualized as the approximate radius of the local window used to compute the means. The eps argument specifies the grid cell size of the output grid. The output is a **spatstat** im object, which can be converted to a **terra** SpatRaster object. The coordinate reference system information is lost in the conversion, so the same CRS as the original abla_ppp object is assigned to the new raster dataset.

```
abla_im <- Smooth.ppp(abla_ppp,
                       sigma=10000,
                       eps=c(1000, 1000))
class(abla_im)
## [1] "im"
abla_grid <- rast(abla_im)
crs(abla_grid) <-"epsg:32610"
```

The output provides a smoothed estimate of the proportion of plots with subalpine fir (Figure 12.3).

```
abla_df <- rasterdf(abla_grid)
ggplot(data = abla_df) +
  geom_raster(aes(x = x,
                  y = y,
                  fill = value)) +
  scale_fill_gradient(name = "Subalpine Fir",
                      low = "lightyellow",
                      high = "darkgreen",
```

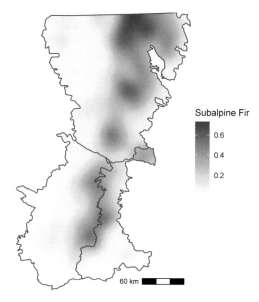

FIGURE 12.3
Smoothed proportion of forest inventory plots with subalpine fir in the Washington Cascades.

```
                    na.value = NA) +
  geom_sf(data = boundary_utm,
          fill = NA) +
  annotation_scale(location = 'br') +
  coord_sf(expand = F) +
  theme_void()
```

The same procedure is repeated for Douglas-fir to generate a smoothed raster of kernel-weighted means.

```
psme_ppp <- as.ppp(psme_utm)
Window(psme_ppp) <- as.owin(boundary_utm)
psme_im <- Smooth.ppp(psme_ppp,
                      sigma=10000,
                      eps=c(1000, 1000))
psme_grid <- rast(psme_im)
crs(psme_grid) <-"epsg:32610"
```

Mapping the smoothed occurrence data helps to show that Douglas-fir is broadly distributed in the southwestern portion of Washington Cascades (Figure 12.4), which is not as clear when looking at the individual points (Figure 12.2).

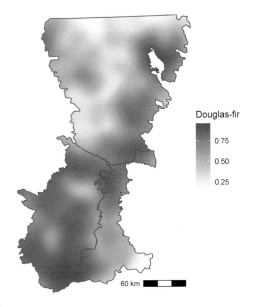

FIGURE 12.4
Smoothed proportion of forest inventory plots with Douglas-fir in the Washington Cascades.

```
psme_df <- rasterdf(psme_grid)
ggplot(data = psme_df) +
  geom_raster(aes(x = x,
                  y = y,
                  fill = value)) +
  scale_fill_gradient(name = "Douglas-fir",
                      low = "lightyellow",
                      high = "darkgreen",
                      na.value = NA) +
  geom_sf(data = boundary_utm,
          fill = NA) +
  annotation_scale(location = 'br') +
  coord_sf(expand = F) +
  theme_void()
```

12.2 WorldClim Historical Climate Data

The WorldClim dataset will be used to analyze relationships between tree species and climate. Worldclim is a downscaled global climate dataset available

TABLE 12.1
Wordclim bioclimatic variable codes and descriptions.

Code	Description
bio1	Annual Mean Temperature
bio2	Mean Diurnal Range
bio3	Isothermality (BIO2/BIO7) ($\times 100$)
bio4	Temperature Seasonality
bio5	Max Temperature of Warmest Month
bio6	Min Temperature of Coldest Month
bio7	Temperature Annual Range
bio8	Mean Temperature of Wettest Quarter
bio9	Mean Temperature of Driest Quarter
bio10	Mean Temperature of Warmest Quarter
bio11	Mean Temperature of Coldest Quarter
bio12	Annual Precipitation
bio13	Precipitation of Wettest Month
bio14	Precipitation of Driest Month
bio15	Precipitation Seasonality
bio16	Precipitation of Wettest Quarter
bio17	Precipitation of Driest Quarter
bio18	Precipitation of Warmest Quarter
bio19	Precipitation of Coldest Quarter

as a 1000 m grid that captures fine-scale variation associated with elevation (Fick and Hijmans, 2017). Thus, it is well-suited for rugged terrain like the Washington Cascades. The basic WorClim variables include monthly temperature and precipitation summarized as monthly climatologies over a 30-year period. These monthly data have been used to generate a set of bioclimatic variables that are more suitable as predictors in species distribution models (Table 12.1).

```
wcbio <- rast("wc2.1_30s_bio_washington.tif")
nlyr(wcbio)
## [1] 19
names(wcbio)
##  [1] "wc2.1_30s_bio_washington_1"
##  [2] "wc2.1_30s_bio_washington_2"
##  [3] "wc2.1_30s_bio_washington_3"
##  [4] "wc2.1_30s_bio_washington_4"
##  [5] "wc2.1_30s_bio_washington_5"
##  [6] "wc2.1_30s_bio_washington_6"
```

```
##  [7] "wc2.1_30s_bio_washington_7"
##  [8] "wc2.1_30s_bio_washington_8"
##  [9] "wc2.1_30s_bio_washington_9"
## [10] "wc2.1_30s_bio_washington_10"
## [11] "wc2.1_30s_bio_washington_11"
## [12] "wc2.1_30s_bio_washington_12"
## [13] "wc2.1_30s_bio_washington_13"
## [14] "wc2.1_30s_bio_washington_14"
## [15] "wc2.1_30s_bio_washington_15"
## [16] "wc2.1_30s_bio_washington_16"
## [17] "wc2.1_30s_bio_washington_17"
## [18] "wc2.1_30s_bio_washington_18"
## [19] "wc2.1_30s_bio_washington_19"
```

The default layer names are replaced by a set of abbreviated codes.

```
wcbnames <- paste0("bio", c(1, 10:19, 2:9))
names(wcbio) <- wcbnames
wcbnames
##  [1] "bio1"  "bio10" "bio11" "bio12" "bio13" "bio14" "bio15"
##  [8] "bio16" "bio17" "bio18" "bio19" "bio2"  "bio3"  "bio4"
## [15] "bio5"  "bio6"  "bio7"  "bio8"  "bio9"
```

The Washington Cascades ecoregion boundaries are reprojected into a geographic coordinate system with WGS84 datum to match the coordinate reference system of the plot data and the WorldClim data. These polygons are then used to crop and mask the Washington Cascades from the larger Worldclim dataset.

```
boundary_wgs84 <- st_transform(wacascades, st_crs(wcbio))
wcbio_crop <- crop(wcbio, vect(boundary_wgs84))
wcbio_msk <- mask(wcbio_crop, vect(boundary_wgs84))
```

The map of maximum temperature during the warmest month of the year highlights the effects of elevation, with the highest temperatures at the fringes of the mountain range and in the larger river valleys. The lowest temperatures occur along the Cascade crest and at the peaks of the large volcanoes (Figure 12.5).

```
mtwm_df <- rasterdf(wcbio_msk[["bio5"]])
ggplot(data = mtwm_df) +
  geom_raster(aes(x = x,
                  y = y,
```

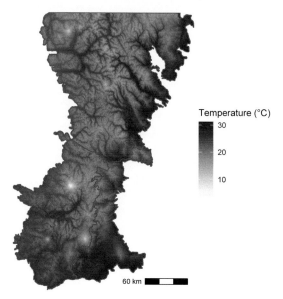

FIGURE 12.5
WorldClim thirty-year climatology of maximum temperature during the
warmest month of the year for the Washington Cascades.

```
                  fill = value)) +
  scale_fill_gradient(name = "Temperature (\u00B0C)",
                      low = "lightyellow",
                      high = "darkred",
                      na.value = NA) +
  geom_sf(data = boundary_wgs84,
          fill = NA) +
  annotation_scale(location = 'br') +
  coord_sf(expand = F) +
  theme_void()
```

The map of precipitation during the wettest month shows the interactions of
moist maritime air with topography, with the highest values occurring west
of the Cascade Crest and the lowest values in the rain shadow on the eastern
slopes (Figure 12.6).

```
mtwm_df <- rasterdf(wcbio_msk[["bio13"]])
ggplot(data = mtwm_df) +
  geom_raster(aes(x = x,
                  y = y,
```

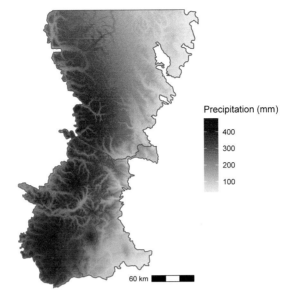

FIGURE 12.6
WorldClim thirty-year climatology of precipitation during the wettest month of the year for the Washington Cascades.

```
                fill = value)) +
  scale_fill_gradient(name = "Precipitation (mm)",
                      low = "lightblue",
                      high = "darkblue",
                      na.value = NA) +
  geom_sf(data = boundary_wgs84,
          fill = NA) +
  annotation_scale(location = 'br') +
  coord_sf(expand = F) +
  theme_void()
```

12.3 Modeling the Climate Niche

12.3.1 Subalpine fir

To prepare the data for modeling, the `extract()` function is used to obtain climate variables from `wcbio_mask` for every plot location in the subalpine fir dataset.

```
abla_bio <- extract(wcbio_msk, vect(abla_pts)) %>%
  bind_cols(abla_pts) %>%
  as.data.frame()
names(abla_bio)
##  [1] "ID"        "bio1"      "bio10"     "bio11"     "bio12"
##  [6] "bio13"     "bio14"     "bio15"     "bio16"     "bio17"
## [11] "bio18"     "bio19"     "bio2"      "bio3"      "bio4"
## [16] "bio5"      "bio6"      "bio7"      "bio8"      "bio9"
## [21] "abla"      "geometry"
```

To assess the predictive accuracy of the model, a subset of the data needs to be held out from the calibration step and reserved for validation. The data can be split using the `sample_frac()` function from the **dplyr** package, where the `size` argument is the proportion of observations in the sample. There is a similar function called `sample_n()` that can be used to sample a specific number of observations. The `anti_join()` function is then used to select the validation points as all the observations in `abla_bio` that are not present in `abla_train`.

```
abla_train <- abla_bio %>%
  sample_frac(size = 0.7)
abla_val <- abla_bio %>%
  anti_join(abla_train, by = "ID")
```

Boosted regression trees are used to model the relationships between presence/absence of subalpine fir and the 19 bioclimatic variables. BRT is a machine learning method that combines regression trees, which relate species occurrence to a set of climate variables using a series of recursive binary splits, with boosting, an ensemble approach that combines many simple models to improve predictive accuracy (Elith et al., 2008). BRT models can account for non-linear relationships with climate as well as interactions between two or more climate variables. In addition to generating predictions, information about the relative importance of climate predictors can be derived, and climate effects can be visualized using partial regression plots.

BRT models are run using the `gbm.step()` function from the **dismo** package. A data frame containing training data is provided along with the indices of the columns to use as the predictor variables (`gbm.x`) and the response variable (`gbm.y`). The `family = "bernoulli"` argument indicates that the model will treat the response as a binomial (presence/absence) variable. The `tree.complexity`, `learning.rate`, and `bag.fraction` arguments control aspects of the gradient boosting algorithm that is used to fit the model. Consult Elith et al. (2008) for more details on these parameters as well as recommended approaches for parameter selection. Here, a set of values that have been determined to yield good predictions with these datasets is used. Other arguments control whether plots and output to the console are automatically produced.

```
set.seed(22003)
abla_mod <- gbm.step(data = abla_train,
                     gbm.x = 2:20,
                     gbm.y = 21,
                     family = "bernoulli",
                     tree.complexity = 3,
                     learning.rate = 0.01,
                     bag.fraction = 0.5,
                     plot.main = FALSE,
                     verbose = FALSE,
                     silent = TRUE)
```

After running the model, the summary function can be used to extract information about the relative importance of the predictor variables. The relative importance is based on the number of times a variable is selected for splitting the regression trees weighted by the resulting improvement to the BRT model. The importance values are then rescaled, so they sum to 100. For subalpine fir, the mean annual temperature (bio1) has a stronger influence in the model than any other bioclimatic variable, followed by the mean temperature of the warmest quarter (bio10) and mean temperature of the driest quarter (bio9).

```
abla_imp <- summary(abla_mod, plotit = FALSE)
abla_imp
##            var   rel.inf
## bio1      bio1 35.991047
## bio10    bio10 10.329919
## bio9      bio9  6.501694
## bio18    bio18  5.563223
## bio14    bio14  5.328320
## bio3      bio3  5.237513
## bio17    bio17  4.825669
## bio15    bio15  4.309203
## bio8      bio8  3.839852
## bio11    bio11  3.187888
## bio5      bio5  2.698535
## bio4      bio4  2.285570
## bio6      bio6  2.238827
## bio12    bio12  1.605638
## bio13    bio13  1.504464
## bio16    bio16  1.463679
## bio2      bio2  1.358105
```

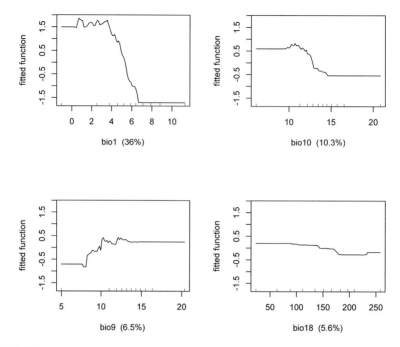

FIGURE 12.7
Partial regression plots showing the response of subalpine fir to WorldClim
bioclimatic indices in a boosted regression tree model.

```
## bio7    bio7   1.002929
## bio19 bio19   0.727925
```

Partial residual plots show the nonparametric relationships between subalpine
fir occurrence and the predictor variables. They can be generated using the
gbm.plot() function from the **dismo** package. Here, the partial plots for the
four most important predictors are shown in a 2 x 2 layout.

Over the range of temperatures in the study area, subalpine fir occurs at the
lowest mean annual temperatures (bio1) and declines monotonically with increasing temperature (Figure 12.7). Subalpine fir occurrence also decreases with
mean temperature of the warmest quarter (bio10), increases with mean temperature of the driest quarter (bio9), and decreases slightly with precipitation
in the warmest quarter (bio18).

```
gbm.plot(abla_mod,
         n.plots = 4,
```

```
        write.title = FALSE,
        plot.layout = c(2, 2))
```

The predicted probability of subalpine fir occurrence can be mapped using the **terra** prediction function. The `SpatRaster` object containing the predictor variables is the first argument, followed by the model object. The `type = "response"` argument is also necessary to transform the predictions to the 0-1 scale, and `na.rm = TRUE` suppresses predictions outside of the study area where there are no data.

```
abla_cur <- predict(object = wcbio_msk,
                    model = abla_mod,
                    type = "response",
                    na.rm = TRUE)
```

Compared to the smoothed map (Figure 12.3), the predicted map based on the WorldClim bioclimatic variables shows much more local variability associated with topography (Figure 12.8). In particular, the species is predicted to be most abundant on high ridges at the Cascade crest and eastward. The circular rings are the upper elevation limits on large volcanoes such as Mount Ranier and Mount Adams.

```
abla_cur_df <- rasterdf(abla_cur)
ggplot(data = abla_cur_df) +
  geom_raster(aes(x = x,
                  y = y,
                  fill = value)) +
  scale_fill_gradient(name = "Subalpine Fir",
                      low = "lightyellow",
                      high = "darkgreen",
                      na.value = NA) +
  geom_sf(data = boundary_wgs84,
          fill = NA) +
  annotation_scale(location = 'br') +
  coord_sf(expand = F) +
  theme_void()
```

12.3.2 Douglas-fir

The data preparation and modeling steps are repeated for Douglas-fir.

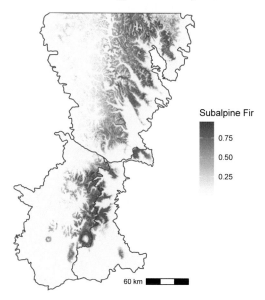

FIGURE 12.8
Predicted probability of subalpine fir occurrence based on historical WorldClim
bioclimatic indices.

```
psme_bio <- extract(wcbio_msk, vect(psme_pts)) %>%
  bind_cols(psme_pts) %>%
  as.data.frame()
psme_train <- psme_bio %>%
  sample_frac(size = 0.7)
psme_val <- psme_bio %>%
  anti_join(abla_train, by = "ID")
set.seed(22004)
psme_mod <- gbm.step(data=psme_train,
                     gbm.x = 2:20,
                     gbm.y = 21,
                     family = "bernoulli",
                     tree.complexity = 3,
                     learning.rate = 0.01,
                     bag.fraction = 0.5,
                     plot.main = FALSE,
                     verbose = FALSE,
                     silent = TRUE)
```

The variable importance ranking is different for Douglas-fir compared to
Subalpine Fir. The most important variables are the maximum temperature

of the warmest month (bio5) and the mean temperature of the warmest quarter (bio10), followed by the precipitation of the driest month (bio14) and precipitation seasonality (bio15).

```
psme_imp <- summary(psme_mod, plotit = FALSE)
psme_imp
##           var    rel.inf
## bio5    bio5 21.510714
## bio10  bio10 12.597744
## bio14  bio14  8.426697
## bio15  bio15  5.857719
## bio3    bio3  5.635875
## bio17  bio17  5.056038
## bio9    bio9  4.955814
## bio18  bio18  4.513356
## bio4    bio4  4.350895
## bio2    bio2  4.116169
## bio12  bio12  4.063634
## bio11  bio11  3.546623
## bio1    bio1  3.225254
## bio7    bio7  2.938707
## bio6    bio6  2.555639
## bio8    bio8  2.435878
## bio16  bio16  1.673572
## bio13  bio13  1.274735
## bio19  bio19  1.264938
```

The relationships between Douglas-fir occurrence and the two temperature variables are unimodal, with species occurrence peaking across a range of optimal temperatures and decreasing at cooler and warmer temperatures (Figure 12.9). Douglas-fir also has a unimodal relationship with precipitation in the driest month (bio14) and an increasing relationship with precipitation seasonality (bio15).

```
gbm.plot(psme_mod,
         n.plots = 4,
         write.title = FALSE,
         plot.layout = c(2, 2))
```

The predicted occurrence map shows that Douglas-fir is widespread throughout the southern part of the Washington Cascades but is confined to lower-elevation valleys in the northern part of the region (Figure 12.10).

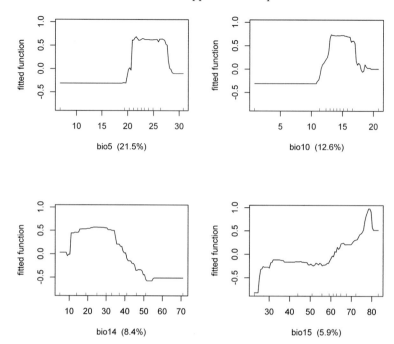

FIGURE 12.9
Partial regression plots showing the response of Douglas-fir to WorldClim
bioclimatic indices in a boosted regression tree model.

```
psme_cur <- predict(object = wcbio_msk,
                    model = psme_mod,
                    type = "response",
                    na.rm = TRUE)
psme_cur_df <- rasterdf(psme_cur)
ggplot(data = psme_cur_df) +
  geom_raster(aes(x = x,
                  y = y,
                  fill = value)) +
  scale_fill_gradient(name = "Douglas-fir",
                      low = "lightyellow",
                      high = "darkgreen",
                      na.value = NA) +
  geom_sf(data = boundary_wgs84,
          fill = NA) +
  annotation_scale(location = 'br') +
  coord_sf(expand = F) +
  theme_void()
```

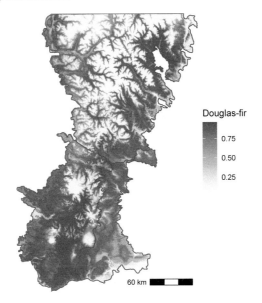

FIGURE 12.10
Predicted probability of Douglas-fir occurrence based on historical WorldClim
bioclimatic indices.

12.4   Accuracy Assessment

To understand how accurate the species distribution model predictions are,
the `predict()` function is used to generate predictions based on the validation
dataset. In this example, the `predict()` method for `gbm` objects is invoked,
requiring different arguments than the previous examples, which used the
`predict()` method for `SpatRaster` objects. The first argument is the `gbm` model,
and the `newdata` argument is used to specify the `abla_val` data frame, which was
created earlier when the full dataset was split into training and validation sets.
The vector of predicted values is combined with the observed values from the
`abla_val` data frame, and the `prediction()` function from the **ROCR** package
is used to generate a `prediction` object that can be used for computing accuracy
statistics.

```
abla_pred <- predict(abla_mod,
                     newdata = abla_val,
                     type = "response")
abla_predobs <- prediction(abla_pred, abla_val$abla)
```

Although the observations are binary, the output of the model is a continuous probability between zero and one. The receiver operating characteristic curve (ROC) is commonly used to evaluate the predictive capability of such a model. The ROC is generated by using the predicted probability to classify presence or absence over a range of cutoff values and exploring the tradeoff between predictions of presence and absence. At the extreme, if all predicted probabilities less than one are assumed to be positive, then all points are classified as present. In this case, the model will correctly predict all the locations where subalpine fir is actually present but fail to predict any of the points where it is actually absent. The reverse is true if zero is used as the threshold and all points are classified as absent.

The ROC curve describes the changes in true positive rates and false positive rates over all possible cutoff values. The `performance()` function in the **ROCR** package is used to generate a `performance` object that includes these two measures.

```
abla_roc = performance(abla_predobs,
                       measure = "tpr",
                       x.measure = "fpr")
class(abla_roc)
## [1] "performance"
## attr(,"package")
## [1] "ROCR"
```

The `performance` object could be visualized using a generic `plot()` method, but this example will show how to extract the underlying data and generate a ROC curve with `ggplot()`. The **ROCR** package uses the S4 object system. Details of S4 are outside the scope of this book, but at the most basic level, these objects require new functions to visualize and extract their components, which are stored in "slots."

```
slotNames(abla_roc)
## [1] "x.name"      "y.name"      "alpha.name"
## [4] "x.values"    "y.values"    "alpha.values"
```

The `slot()` function can be used to access the slots in an S4 object. The `x.values` and `y.values` contain the false positive rates and the true positive rates for the ROC curve. These data are stored as lists with a single element. Double brackets are used to subscript these list elements and return numeric vectors, which are combined into a data frame.

```
abla_fpr <- slot(abla_roc, "x.values")[[1]]
```

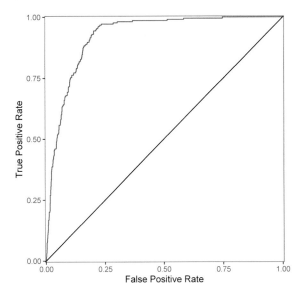

FIGURE 12.11
Area under the receiver operating characteristics (ROC) curve for predictions
of subalpine fir occurrence based on WorldClim bioclimatic indices.

```
abla_tpr <- slot(abla_roc, "y.values")[[1]]
abla_aucplot <- data.frame(abla_fpr, abla_tpr)
```

This data frame can be used with `ggplot()` to graph the ROC curve. As the
false positive rate (equal to one minus the true negative rate) decreases, the
true positive rate also decreases (Figure 12.11).

```
ggplot(data = abla_aucplot) +
  geom_line(aes(x = abla_fpr,
                y = abla_tpr),
            col = "red") +
  labs(x = "False Positive Rate",
       y = "True Positive Rate") +
  geom_abline(slope = 1, intercept = 0) +
  scale_x_continuous(expand = c(0.005, 0)) +
  scale_y_continuous(expand = c(0.005, 0)) +
  coord_fixed() +
  theme_bw()
```

Another way to visualize these relationships is to graph the overall accuracy,
true positive rate, and true negative rate as a function of the cutoff used for

prediction. The `performance()` function is used to create `performance` objects containing these three statistics. The values are then extracted and combined into a long-format data frame.

```
abla_all = performance(abla_predobs,
                        measure = "acc")
abla_pos = performance(abla_predobs,
                        measure = "tpr")
abla_neg = performance(abla_predobs,
                        measure = "tnr")
cutoff <- slot(abla_all, "x.values")[[1]]
totacc <- slot(abla_all, "y.values")[[1]]
posacc <- slot(abla_pos, "y.values")[[1]]
negacc <- slot(abla_neg, "y.values")[[1]]
abla_accplot <- data.frame(cutoff,
                           totacc,
                           posacc,
                           negacc) %>%
  pivot_longer(cols = one_of("totacc",
                             "posacc",
                             "negacc"),
               values_to = "accval",
               names_to = "accstat")
```

The results suggest that a cutoff value around 0.25 would be effective for classifying presence or absence based on the predictions (Figure 12.12). The overall accuracy is close to the maximum, and the true negative and true positive rates are balanced and relatively high.

```
ggplot(data = abla_accplot) +
  geom_line(aes(x = cutoff,
                y = accval,
                col = accstat)) +
    labs(x = "Classification Cutoff",
         y = "Classification Accuracy",
         color = "Accuracy Statistic") +
    scale_color_discrete(labels = c("True Negative Rate",
                                    "True Positive Rate",
                                    "Overall Accuracy")) +
  scale_x_continuous(expand = c(0.005, 0)) +
  scale_y_continuous(expand = c(0.005, 0)) +
  coord_fixed() +
  theme_bw()
```

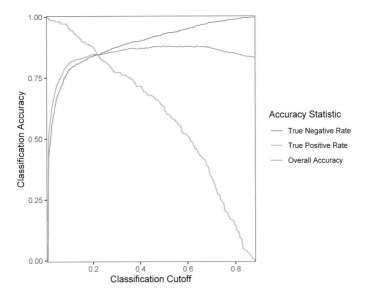

FIGURE 12.12
Overall accuracy, true positive rate, and true negative rate for predictions of subalpine fir occurrence based on WorldClim bioclimatic indices.

The area under the ROC curve, often abbreviated as the AUROC or just the AUC, is frequently used as an index of the predictive power of a probabilistic model for binary data like species presence/absence. Values range from a minimum of 0.5 to a maximum of 1.0. The AUC statistic can be calculated using the performance() function.

```
abla_aucval <- performance(abla_predobs, measure = "auc")
slot(abla_aucval, "y.values")[[1]]
## [1] 0.9231791
```

The AUC for the subalpine fir model is 0.92, which is relatively high value and indicates that predictions are accurate over a range of cutoff values.

12.5 Climate Change Projections

The WorldClim dataset also includes projected future climate grids based on the outputs of general circulation models (GCMs) from the Coupled Model Intercomparison Project (CMIP) version 6. This example uses a projection for 2061-2080 from the Max Planck Institute Earth System Model

(MPI-ESM1.2). It is based on RCP4.5 global forcing pathway, which assumes that carbon dioxide (CO2) emissions will start to decline before 2045 and reach approximately half of their 2050 levels by 2100. The variables are the same bioclimatic indices that were computed for historical climatology. The data are read into a raster object, assigned layer names, and cropped and masked to the Washington Cascades study area.

```
wcproj <- rast("wc2.1_30s_bioc_MPI-ESM1-2-HR_ssp245_2061-2080_wa.tif")
nlyr(wcproj)
## [1] 19
wcbnames <- paste0("bio", 1:19)
names(wcproj) <- wcbnames
wcproj_crop <- crop(wcproj, vect(boundary_wgs84))
wcproj_msk <- mask(wcproj_crop, vect(boundary_wgs84))
```

The **terra** predict() function is used to generate predictions of subalpine fir distribution under the future climate scenario. The object argument is used to specify the raster with projected climate variables. This raster must have the same layer names as the original variables that were used to train the model. The output, abla_proj, is combined with the predictions under current conditions, abla_cur, to create a multi-layer raster.

```
abla_proj <- predict(object = wcproj_msk,
                     model = abla_mod,
                     type = "response",
                     na.rm = TRUE)
abla_chg <- c(abla_cur, abla_proj)
names(abla_chg) <- c("Current", "Future")
```

To display the predictions as binary presence or absence outcomes, the probabilities are classified using a cutoff value of 0.25.

```
abla_clas <- ifel(abla_chg > 0.25, 1, 0)
```

Comparing these two maps shows that the projected range of subalpine fir is considerably smaller under a future, warmer climate, with the species restricted to the highest elevations in the Washington Cascades (Figure 12.13).

```
abla_clas_df <- rasterdf(abla_clas)
ggplot(data = abla_clas_df) +
  geom_raster(aes(x = x,
                  y = y,
                  fill = as.character(value))) +
```

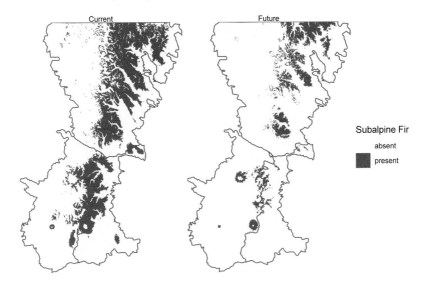

FIGURE 12.13
Predicted probability of subalpine fir occurrence based on future (2061-2080)
WorldClim bioclimatic indices.

```
scale_fill_manual(name = "Subalpine Fir",
                  values = c("lightyellow",
                             "darkgreen"),
                  labels = c("absent",
                             "present"),
                  na.translate = FALSE) +
facet_wrap(~ variable) +
geom_sf(data = boundary_wgs84,
        fill = NA) +
coord_sf(expand = F) +
theme_void()
```

In addition to generating maps, the data frame generated by the `rasterdf()`
function can also be used to create other types of graphs based on the values
in a raster dataset. In this example, histograms of current and future mean
annual temperatures show that the distribution of mean annual temperatures
will be considerably higher under the future climate scenario (Figure 12.14).

```
annmean <- c(wcbio_msk[["bio1"]], wcproj_msk[["bio1"]])
names(annmean) <- c("Current", "Future")
annmean_df <- rasterdf(annmean)
```

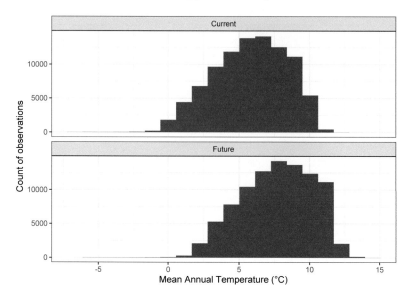

FIGURE 12.14
Histogram of current and projected future mean annual temperatures for the
Washington Cascades.

```
ggplot(data = annmean_df) +
  geom_histogram(aes(x = value), bins = 20) +
  labs(x = "Mean Annual Temperature (\u00B0C)",
       y = "Count of observations") +
  facet_wrap(~ variable, ncol = 1) +
  theme_bw()
```

12.6   Practice

1. Carry out an accurate assessment of the species distribution model
   for Douglas-fir and compare the results to those obtained for subalpine fir.

2. Predict the future distribution of Douglas-fir using the climate projections and compare the results with the projected future distribution
   of subalpine fir.

3. Generate a map that shows where the distribution of subalpine fir is projected to expand, contract, and remain the same under future climate projections. Then generate a similar map for Douglas-fir and compare the two.

Appendix

Required Software

Running the examples and practice exercises in this book requires installation of the R software environment if it is not already on your computer. RStudio, an integrated development environment (IDE) that makes R easier to use, is also highly recommended. R and RStudio are both free and open source, which means that the software is available to all users at no cost and can be redistributed and modified.

R

First, download and install R from CRAN, the Comprehensive R Archive Network.

Go to `https://cloud.r-project.org/` and click **Download R for Windows**, then under "Subdirectories" on the next page click **base**. Finally, click **Download R "X.X.X" for Windows**, where "X.X.X" is the current R version.

To install R on a Mac, choose **Download R for (Mac) OS X** and click **R-X.X.X.pkg** on the next page to download the installer.

After downloading the installer, click it to begin the installation process. If you are prompted with the message "The publisher could not be verified. Are you sure you want to run this software?", then click **Run**. Next, select English as the language you want R to use. As you proceed through the installation screens, it is recommended to leave all options on their default settings.

RStudio

RStudio is an integrated development environment (IDE) for R. After you have installed R, download and install the free version of RStudio Desktop from `http://www.rstudio.com/download`. Click the link to download the free version of RStudio Desktop. On the next page, click the button to download RStudio for Windows. This page also contains links to download RStudio installers for Mac and Linux systems.

After downloading the installer, click it to begin the installation process. If you are prompted with the message "Do you want to run this file?" then click **Run**. As you proceed through the installation screens, it is recommended to leave all options on their default settings.

If you already have RStudio installed on your computer, you can update it to the newest version by following the directions above or by running RStudio and going to **Help > Check for Updates**.

When you start RStudio, R will also start automatically and run within the RStudio interface.

Packages

Throughout the class, you will need to install and use a variety of R packages. An R **package** is a collection of functions, data, and documentation that extends the capabilities of base R. In this class, we will use a number of classes that provide functions for reading and processing geospatial datasets, implementing various spatial analysis techniques, and visualizing the results of these analyses.

You can install a package with a single line of code:

```
install.packages("ggplot2")
```

R will download the packages from CRAN and install them onto your computer. If you have problems installing, make sure that you are connected to the internet and that https://cloud.r-project.org/ isn't blocked by your firewall or proxy. Note that in RStudio, you can also search for and install packages by selecting **Tools > Install Packages...** from the menu.

You will not be able to use the functions, objects, and help files in a package until you load it. Once you have installed a package, you can load it with the library() function:

```
library(ggplot2)
```

The messages tell you that R is loading the ggplot2 package, which we will use in one of the first labs.

Maintaining the Software

The main R installation, as well as most R packages are updated frequently, with updates occurring several times a year. It is important to keep your software up to date to be sure that it is bug-free and that you are working with the most recent versions of critical packages.

Updating R

The simplest way to update R is by going to `https://cloud.r-project.org/`, and downloading and installing the latest version of R if it is newer than the version currently on your computer. It is usually important to update following a major version change (e.g., a change from R version 4.0.3 to version 4.1.1). However, it may be less critical to update following a minor version change (e.g., from R version 4.0.3 to 4.0.4).

You can also use the `updateR()` function in the **installr**(Galili, 2021) package to update R automatically. To update using this function, you should run it from the R GUI, not in RStudio. The function offers some handy options, including an option to copy the R packages from the library of your existing R installation to the new one. However, this option does not always work correctly. It is often more straightforward just to reinstall any packages that you need after updating your R installation.

Updating packages

In some cases, you may need to update one or more of your packages to a later version without installing a new version of R. You can accomplish this task with the `update.packages()` function. The following function will display each package on the screen and prompt the user to select yes (y), no (N), or cancel (c).

```
update.packages()
```

Specifying the `ask = FALSE` argument will automatically update all packages without prompting the user.

```
update.packages(ask = FALSE)
```

Note that in RStudio, you can also update packages by selecting **Tools > Check for Package Updates...** from the menu. This approach is particularly handy if you just need to update one or a few packages.

Updating packages after updating R

Unfortunately, there is no straightforward way to transfer all of your packages from an old version of R to a new version of R. As mentioned earlier, the `updateR()` function in the **installr** package can sometimes copy and update your packages automatically when you install a new R version, but it doesn't always work depending on where your packages are stored and how the permissions on your computer are set up. Another option is to simply reinstall all your packages in the new R installation. However, this can be time-consuming to do manually. The following code uses the `installed.packages()` function to extract package information and automatically install the packages in a new version of R.

Run the following code in your old version of R.

```
# Store information about installed packages in a data frame
mypackages <- as.data.frame(installed.packages())
# Explore the data frame if you wish
View(mypackages)
# Save the data to a comma delimited text file
write.csv(mypackages, 'old_packages.csv')
```

Then close the old version of R, open the new version, copy `old_packages.csv` into the working directory, and run the following code.

```
# Read in the save list of old package
oldpackages <- read.csv('old_packages.csv')
# Read in the list of base R packages in the new version
curpackages <- as.data.frame(installed.packages())
# Generate a vector of add-on packages to be installed
newpackages <- setdiff(oldpackages$Package, curpackages$Package)
# Install the packages
install.packages(newpackages)
```

Updating RStudio

RStudio should also be updated periodically. To see if there is a new version, go to **Help > Check for Updates** in the RStudio menu. If a new version of RStudio is available, then follow the instructions to download and install it.

Managing RStudio Projects

When using R and RStudio, you will end up working with a variety of different computer files. RStudio allows users to create *projects* to help manage all the files associated with a particular workflow in a single folder. These include:

- RStudio project file (.RProj)

- Workspace file (.RData)

- History file (.Rhistory)

- Script files (.R and .Rmd files)

- Input data files (possibly including .csv and .xlsx files for tabular data, .tif files for gridded data, and ESRI shapefiles for vector data)

- Output files (possibility including all input data file formats, .jpg, .png, or .tif files for graphics, and .pdf or .html files for formatted reports)

Setting up RStudio projects

The recommended approach for setting up an RStudio project involves creating a folder for the project and then saving all project files in that folder. This is a relatively simple approach that has the advantage of being totally self-contained. To move or copy a project, all you need to do is move or copy the folder and everything will still work. You do not need to specify directory paths in your code - by default, R input and output functions will work with files in the main project directory. Eventually, you may need to develop more complex scripts that specify explicit paths to other directories in your file system. However, this simpler method is highly recommended for those learning R and RStudio.

The recommended steps are as follows:

1. Start by creating a new folder for the project.

2. In the RStudio menu bar, go to File>Project and select Existing Directory in the Create Project box (sometimes, it takes a while for this box to pop up).

3. Navigate to the folder that you just created and select Create Project.

4. The folder in which the RStudio project was created should contain the following items:

- A hidden .Rproj.user folder that you don't need to worry about.

- An .RData file that contains the saved R workspace.

- An .Rhistory file that contains the history of all the code that has been executed in the project.

- The R Project file - DemoProject.Rproj in this example.

5. To open an RStudio project, you can do one of the following:

- Select File>Open Project from the RStudio menu bar, navigate to the project directory, and select the .Rproj file.

- Navigate to the project directory in RStudio and double-click on the .Rproj file.

Using RStudio projects

The folder in which the RStudio project was created also serves as the R *working directory*. The .RData and .Rhistory files will be saved here by default. When data are imported, R will automatically look for the input data in the working directory unless a different path is specified. When data are exported, R will automatically put the output in the working directory unless a different path is specified.

The most critical components of your projects are your R script files and your input data. If you have these files, then you can always run your code again to generate your outputs. You should save your R scripts files frequently while you are working, and it is also advisable to save backup copies before making major changes. When you quit RStudio, you will typically see a prompt that shows you any unsaved files and asks if you want to save them.

It is usually a good idea to save any unsaved script files so that you don't lose your most recent work. However, in most cases, it is better to *not* save the workspace image (.RData) file. Instead, you can just re-run your script and regenerate the workspace the next time you open the project. Using this approach, you can keep the focus on maintaining your code instead of trying to keep track of all the R objects that are generated when the code runs.

──────────────────────────

Package Conflicts

One of the trickiest challenges in working with R is dealing with conflicts between packages that have the same function names. This issue can result in strange errors that are very difficult to diagnose. Consider the following example. We start by loading **tidyr**.

```
library(tidyr)
```

The **tidyr** package has a handy function called `extract()` that splits a data frame column into multiple columns based on a regular expression. This example splits a column of strings into the values before and after the dash.

```
df <- data.frame(x = c(NA, "a-b", "a-d", "b-c", "d-e"))
df
##       x
```

```
## 1 <NA>
## 2  a-b
## 3  a-d
## 4  b-c
## 5  d-e
extract(df, x, c("A", "B"), "([[:alnum:]]+)-([[:alnum:]]+)")
##      A    B
## 1 <NA> <NA>
## 2    a    b
## 3    a    d
## 4    b    c
## 5    d    e
```

But perhaps we also need to load the **terra** package to analyze some raster data.

```
library(terra)
## terra 1.5.34
##
## Attaching package: 'terra'
## The following object is masked from 'package:tidyr':
##
##     extract
```

Now, the `extract()` function returns an error.

```
extract(df, x, c("A", "B"), "([[:alnum:]]+)-([[:alnum:]]+)")
```

What is happening here? After the packages have been loaded, they are visible in the list of attached packages and objects, which can be viewed with the `search()` function.

```
search()
##  [1] ".GlobalEnv"        "package:terra"
##  [3] "package:tidyr"     "package:ggplot2"
##  [5] "package:stats"     "package:graphics"
##  [7] "package:grDevices" "package:utils"
##  [9] "package:datasets"  "package:methods"
## [11] "Autoloads"         "package:base"
```

When a function is called, R goes through all available packages in memory to find one that contains the function. If there are functions with the same name in more than one package, then R will run the function from the first package found in the search list. The other functions are "masked," meaning they are not called by default. The **tidyr** package has an `extract()` function, but so does **terra**. If **terra** comes before **tidyr** in the search list, then the **terra** `extract()` function will be run, and the **tidyr** `extract()` function will be masked.

If you want to choose a function from a particular library, you can call it explicitly using the double-colon `::` operator, e.g., `tidyr::extract()` or `terra::extract()`. Note that the order of packages in the search list is the opposite of the order that they are loaded - the most recently loaded packages mask previously loaded packages.

```
tidyr::extract(df, x, c("A", "B"), "([[:alnum:]]+)-([[:alnum:]]+)")
##        A     B
## 1  <NA>  <NA>
## 2     a     b
## 3     a     d
## 4     b     c
## 5     d     e
```

These function conflicts are a common source of errors in R programming. One way to minimize them is to load your most important packages last instead of first. Also, if you are using a function with a generic name like `extract()` that is found in multiple packages, it is good practice to call it explicitly with the `::` operator. To see if a particular function is present in multiple packages, you can use the `help()` function with the package name as an argument. If that function is present in two or more loaded packages, RStudio will list them in the Help window. Try this out with `help(select)`. You can also look for messages about 'masked' packages that are returned after loadings packages with the `library()` function.

Bibliography

Baddeley, A., Turner, R., and Rubak, E. (2022). *spatstat: Spatial Point Pattern Analysis, Model-Fitting, Simulation, Tests.* R package version 2.3-4.

Bolstad, P. (2019). *GIS Fundamentals: A First Text on Geographic Information Systems, Sixth Edition.* XanEdu Publishing Inc.

Boryan, C., Yang, Z., Mueller, R., and Craig, M. (2011). Monitoring U.S. Agriculture: the U.S. Department of Agriculture, National Agricultural Statistics Service, Cropland Data Layer program. *Geocarto International*, 26(5):341–358.

Breheny, P. and Burchett, W. (2020). *visreg: Visualization of Regression Models.* R package version 2.7.0.

Charney, N. D., Record, S., Gerstner, B. E., Merow, C., Zarnetske, P. L., and Enquist, B. J. (2021). A test of species distribution model transferability across environmental and geographic space for 108 western North American tree species. *Frontiers in Ecology and Evolution*, 9:689295.

Chen, B. (2021). *CropScapeR: Access Cropland Data Layer Data via the CropScape Web Service.* R package version 1.1.3.

Cheng, J., Karambelkar, B., and Xie, Y. (2022). *leaflet: Create Interactive Web Maps with the JavaScript Leaflet Library.* R package version 2.1.1.

Daly, C., Gibson, W. P., Taylor, G. H., Johnson, G. L., and Pasteris, P. (2002). A knowledge-based approach to the statistical mapping of climate. *Climate Research*, 22(2):99–113.

Dark, S. J. and Bram, D. (2007). The modifiable areal unit problem (MAUP) in physical geography. *Progress in Physical Geography*, 31(5):471–479.

Davies, T. M. (2016). *The Book of R: a First Course in Programming and Statistics.* No Starch Press.

Dunnington, D. (2022). *ggspatial: Spatial Data Framework for ggplot2.* R package version 1.1.6.

Edmund, H. and Bell, K. (2020). *prism: Access Data from the Oregon State Prism Climate Project.* R package version 0.2.0.

Eidenshink, J., Schwind, B., Brewer, K., Zhu, Z.-L., Quayle, B., and Howard, S. (2007). A project for monitoring trends in burn severity. *Fire Ecology*, 3(1):3–21.

Elith, J. and Leathwick, J. R. (2009). Species distribution models: ecological explanation and prediction across space and time. *Annual Review of Ecology, Evolution and Systematics*, 40(1):677–697.

Elith, J., Leathwick, J. R., and Hastie, T. (2008). A working guide to boosted regression trees. *Journal of Animal Ecology*, 77(4):802–813.

Fick, S. E. and Hijmans, R. J. (2017). WorldClim 2: new 1-km spatial resolution climate surfaces for global land areas. *International Journal of Climatology*, 37(12):4302–4315.

Galili, T. (2021). *installr: Using R to Install Stuff on Windows OS (Such As: R, Rtools, RStudio, Git, and More!)*. R package version 0.23.2.

Harrower, M. and Brewer, C. A. (2003). ColorBrewer.org: An online tool for selecting colour schemes for maps. *The Cartographic Journal*, 40:27–37.

Hijmans, R. J. (2022). *terra: Spatial Data Analysis*. R package version 1.5-34.

Hijmans, R. J., Phillips, S., Leathwick, J., and Elith, J. (2022). *dismo: Species Distribution Modeling*. R package version 1.3-8.

Kabacoff, R. I. (2015). *R in Action: Data Analysis and Graphics With R*. Simon and Schuster.

Lovelace, R., Nowosad, J., and Muenchow, J. (2019). *Geocomputation With R*. CRC Press.

Miller, J. D. and Thode, A. E. (2007). Quantifying burn severity in a heterogeneous landscape with a relative version of the delta Normalized Burn Ratio (dNBR). *Remote Sensing of Environment*, 109(1):66–80.

Pebesma, E. (2022). *sf: Simple Features for R*. R package version 1.0-7.

Pebesma, E., Mailund, T., Kalinowski, T., and Ucar, I. (2022). *units: Measurement Units for R Vectors*. R package version 0.8-0.

Perpinan Lamigueiro, O. and Hijmans, R. (2022). *rasterVis: Visualization Methods for Raster Data*. R package version 0.51.2.

Radeloff, V. C., Hammer, R. B., Stewart, S. I., Fried, J. S., Holcomb, S. S., and McKeefry, J. F. (2005). The wildland–urban interface in the United States. *Ecological Applications*, 15(3):799–805.

Sing, T., Sander, O., Beerenwinkel, N., and Lengauer, T. (2005). ROCR: visualizing classifier performance in R. *Bioinformatics*, 21(20):7881.

Spinu, V., Grolemund, G., and Wickham, H. (2021). *lubridate: Make Dealing with Dates a Little Easier*. R package version 1.8.0.

Stachelek, J. (2022). *nhdR: Tools for Working with the National Hydrography Dataset*. R package version 0.5.9.

Tennekes, M. (2018). tmap: Thematic maps in R. *Journal of Statistical Software*, 84(6):1–39.

Walker, K. (2022). *tigris: Load Census TIGER/Line Shapefiles*. R package version 1.6.1.

Wickham, H. (2016). *ggplot2: Elegant Graphics for Data Analysis*. Springer.

Wickham, H. (2021). *tidyverse: Easily Install and Load the Tidyverse*. R package version 1.3.1.

Wickham, H., Chang, W., Henry, L., Pedersen, T. L., Takahashi, K., Wilke, C., Woo, K., Yutani, H., and Dunnington, D. (2022a). *ggplot2: Create Elegant Data Visualisations Using the Grammar of Graphics*. R package version 3.3.6.

Wickham, H., François, R., Henry, L., and Müller, K. (2022b). *dplyr: A Grammar of Data Manipulation*. R package version 1.0.9.

Wickham, H. and Girlich, M. (2022). *tidyr: Tidy Messy Data*. R package version 1.2.0.

Wickham, H. and Grolemund, G. (2016). *R for Data Science: Import, Tidy, Transform, Visualize, and Model Data*. O'Reilly Media, Inc.

Wilke, C. O. (2020). *cowplot: Streamlined Plot Theme and Plot Annotations for ggplot2*. R package version 1.1.1.

Wimberly, M. C., Cochrane, M. A., Baer, A. D., and Pabst, K. (2009). Assessing fuel treatment effectiveness using satellite imagery and spatial statistics. *Ecological Applications*, 19:1377–1384.

Wimberly, M. C., Janssen, L. L., Hennessy, D. A., Luri, M., Chowdhury, N. M., and Feng, H. (2017). Cropland expansion and grassland loss in the eastern Dakotas: New insights from a farm-level survey. *Land Use Policy*, 63:160–173.

Wood, S. (2022). *mgcv: Mixed GAM Computation Vehicle with Automatic Smoothness Estimation*. R package version 1.8-40.

Index